Lecture Notes in Networks and Systems

Volume 58

The series "Lecture Notes in Networks and Systems" publishes the latest developments in Networks and Systems—quickly, informally and with high quality. Original research reported in proceedings and post-proceedings represents the core of LNNS.

Volumes published in LNNS embrace all aspects and subfields of, as well as new challenges in, Networks and Systems.

The series contains proceedings and edited volumes in systems and networks, spanning the areas of Cyber-Physical Systems, Autonomous Systems, Sensor Networks, Control Systems, Energy Systems, Automotive Systems, Biological Systems, Vehicular Networking and Connected Vehicles, Aerospace Systems, Automation, Manufacturing, Smart Grids, Nonlinear Systems, Power Systems, Robotics, Social Systems, Economic Systems and other. Of particular value to both the contributors and the readership are the short publication timeframe and the world-wide distribution and exposure which enable both a wide and rapid dissemination of research output.

The series covers the theory, applications, and perspectives on the state of the art and future developments relevant to systems and networks, decision making, control, complex processes and related areas, as embedded in the fields of interdisciplinary and applied sciences, engineering, computer science, physics, economics, social, and life sciences, as well as the paradigms and methodologies behind them.

**** Indexing: The books of this series are submitted to ISI Proceedings, SCOPUS, Google Scholar and Springerlink ****

More information about this series at http://www.springer.com/series/15179

Moses Eterigho Emetere · Esther Titilayo Akinlabi

Introduction to Environmental Data Analysis and Modeling

Springer

Moses Eterigho Emetere
Department of Physics
Covenant University
Ota, Nigeria

Department of Mechanical
Engineering Science
University of Johannesburg
Auckland Park Campus
Johannesburg, South Africa

Esther Titilayo Akinlabi
Department of Mechanical
Engineering Science
University of Johannesburg
Auckland Park Campus
Johannesburg, South Africa

Department of Mechanical Engineering
Covenant University
Ota, Nigeria

ISSN 2367-3370 ISSN 2367-3389 (electronic)
Lecture Notes in Networks and Systems
ISBN 978-3-030-36206-5 ISBN 978-3-030-36207-2 (eBook)
https://doi.org/10.1007/978-3-030-36207-2

This Springer imprint is published by the registered company Springer Nature Switzerland AG
The registered company address is: Gewerbestrasse 11, 6330 Cham, Switzerland

Foreword

Environmental modeling is not just broad but dynamic. The concept of this book is both a guide to a new beginner in the field and a veritable tool for professionals. Authors of the book have deliberated on the subject matter right from the fundamental concepts through methods for numerical analyses and then unto the applications for the treatment of environmental data. It is my delight, therefore, to foreword this book as a major contribution to knowledge for beginner and practitioners of environmental data analyses.

Joshua Olusegun Okeniyi, Ph.D.
Covenant University
Ota, Nigeria

Preface

There is still a significant void in relating huge environmental dataset to emerging mathematical or computational models. This book considers the theoretical background of some special cases with the primary aim of introducing numerical methods for processing dataset, which may be of any form. Computational resolution of environmental dataset alongside the use of open-source libraries was adequately illustrated.

The purpose of writing this book is to solve the challenges of misrepresentation of dataset that are relevant directly or indirectly to the research. This book illustrates new ways of screening dataset or images for maximum utilization. The adoption of various numerical methods in dataset treatment would certainly create a new scientific approach. The focus of this book is to give a new direction on how to analyze environmental measurements to ensure 100% utilization.

This book aims at introducing new ways of data treatment that is based on sound mathematical and computational approach. Most environmental researchers are in the dilemma of working on doctored or prorated dataset from primary sources. Most time, the processes at which the dataset at the primary sources are obtained are unclear or classified. Hence, researchers working on such dataset are likely to be working with high error margin when analyzing the huge volume of the dataset.

This book solves the problem of validating dataset through unique data treatment processes. The book is recommended for research assistants and professionals in the fields of Environmental Sciences, Civil Engineering, Chemical Engineering, Environmental Chemistry, Atmospheric Physics, Space Physics, Environmental Biology, and Communication/Electrical Engineering. In addition, this book is specifically recommended for researchers and scientists in communication and satellite industries, environmental monitoring centers/institutes, and environmental companies or agencies.

Ota, Nigeria/Johannesburg, South Africa — Moses Eterigho Emetere, Ph.D.
Johannesburg, South Africa/Ota, Nigeria — Esther Titilayo Akinlabi, D.Tech.

Contents

Chapter 1
Introduction to Environmental Modeling

From elementary definition, the environment is everything that is around us. The environment are controlled by physical, chemical and other natural forces that enables living things to interact with themselves and other non-living things in a specific way. In order for some living things to adapt to their environment, they migrate, hibernate, shed scale or fur, grow extra layer of scaly skin, or shed leaves etc. In other words, living things interact with their immediate environment to adapt themselves to specific conditions in their environment. In a way, the interaction of living and non-living things in the bid to adapt to physical, chemical and natural environmental forces, inspired civilization and hazards. For example, the changing regional climate system led to the different types of seasonal clothing in different parts of the globe. Also, since the time human embraced industrialization, there are different environmental disasters. The main environmental disaster is pollution. Pollution may be air, land, water or noise. Air pollution has the highest potential to cause death.

Air pollution is the emission of anthropogenic or harmful substances in form of particulates (small-suspended particles of varying sizes) or gas (sulphur dioxide (SO_2), nitrogen oxides (NO_x), ozone (O_3), volatile organic compounds) to the atmosphere [1].

World Health Organization (WHO) reported 4.2 million deaths every year as a result of exposure to ambient (outdoor) air pollution; 3.8 million deaths every year as a result of household exposure to smoke from dirty cook stoves and fuels; and 91% of the world's population lives in places where air quality exceeds WHO guideline limits [2]. Ritchie and Roser [3] reported that the dominant part of pollution related deaths are in Asia—South, Southeast and East Asia alone represented almost 3 million out of 7 million. Annually, pollution related deaths are dominated by China and India—each with between 1.1 and 1.2 million deaths in 2016. The global air pollution chart is shown in Fig. 1.1.

In 2018, WHO made a sterling revelation on the state of water pollution and its extensive hazardous effects that 2.1 billion people are exposed to the hazards of water pollution [4]. For example, 1.3 billion people migrate a round trip of 30 min to get improved water source; 263 million people have difficulties in collecting water at

M. E. Emetere and E. T. Akinlabi, *Introduction to Environmental Data Analysis and Modeling*, Lecture Notes in Networks and Systems 58,
https://doi.org/10.1007/978-3-030-36207-2_1

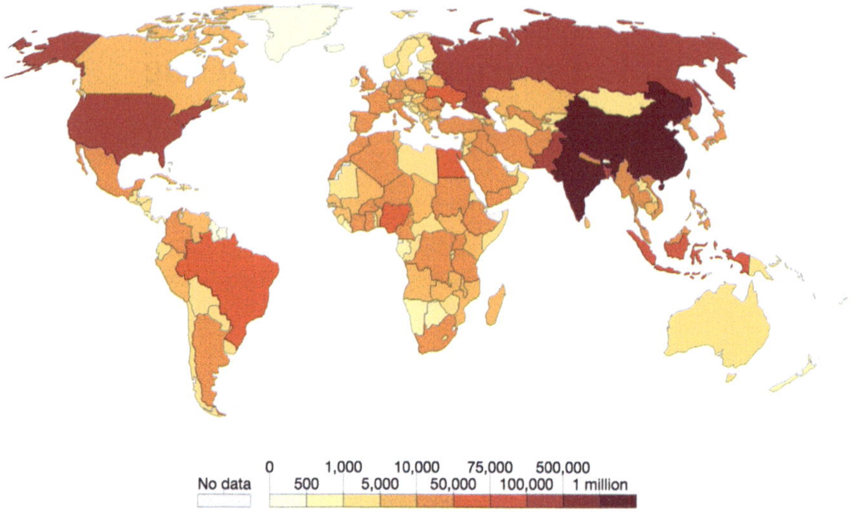

Fig. 1.1 Global air pollution distribution [3]

limited water services; 423 million individuals take water from unprotected wells and springs; and 159 million individuals gathering untreated surface water from lakes, ponds, waterways and streams. Aside the man-made hazards (water and air pollution), there are natural hazards e.g. earthquakes, volcanic eruptions, tremor, wind storm, Tsunami's and different flooding systems. Guha-Sapir et al. [5] gave an in-depth analysis of disaster occurrences across continents in 2016. By number, about 157 disasters were reported in 2016. East Asia was the most hit (54 disasters), and South Asia (40), East Africa (29 disasters), North America (28 disasters), South America (23 disasters) and other regions suffered less than 20 disasters. By percentage, Asia had the highest hit (46.7%), followed by the Americas (24.3%), Africa (16.9%), Europe (8.2%), and Oceania (3.8%). From the above information, the adaptation of living things is quite difficult. The global deaths by natural disaster are shown in Fig. 1.2.

Environmental hazards are associated with causative forces or medium. For example, earthquakes occur when rocks in the earth crust suddenly breaks along a fault to release energy that initiates seismic waves that make the ground shake [7, 8]. Flooding in most parts of the world are naturally caused by heavy rain, overflowing rivers, storm surges and tsunamis, lack of vegetation and melting snow and ice. These factors have been related to some atmospheric effects [8–10]. The causes of volcanic eruption can be tied to the buoyancy of the magma, the pressure from the entrapped gases in the magma and the injection of a new batch of magma into an already filled magma chamber. Windstorm is caused by the gust front (the boundary between descending cold air and warm air at the surface) of an approaching thunderstorm.

One of the main reasons for environmental studies is the quest of human to understand the environment with the singular focus of adaptation. Environmental

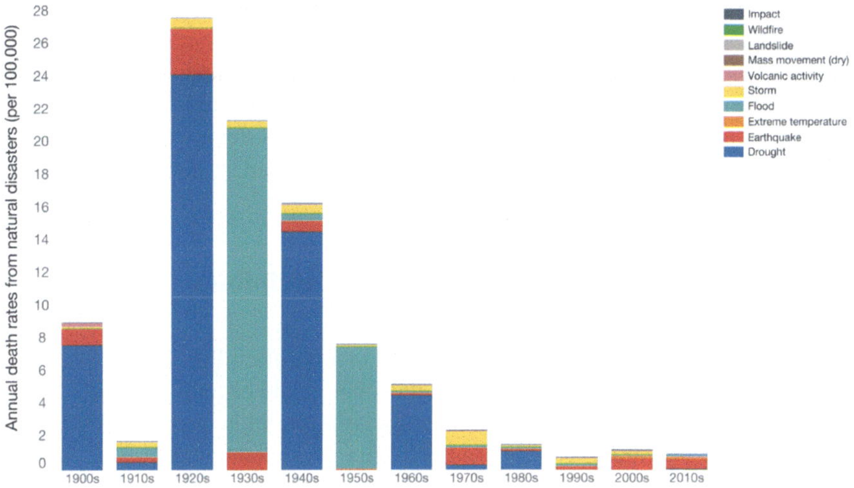

Fig. 1.2 Global deaths by natural disaster [6]

studies have different nomenclature, i.e. environmental science, environmental technology, environmental engineering, environmental health etc. The modes of conducting experiment in environmental studies differ with respect to different disciplines. The different known disciplines in environmental studies are atmospheric physics, meteorology, astronomy, environmental biology, environmental economics, environmental accountancy, environmental chemistry, civil engineering, zoology, ecology etc. All the environmental discipline ensures the sustainability of life form on earth. The dynamism of the forces that controls environment may be complex most time because of its unknown trigger source and the inter-dependency of associated forces. For example, the updraft forces that lift water molecules into the atmosphere are also responsible for aerosol loading. The forces of wind or front have its turn on the direction of this particulates which in the long run leads to heavy down-pour or famine. Hence, the question scientists ask is: ‘how do environmental forces behave?’ The proper documentation of the behaviour of any environmental forces is referred to a model that describes the processes that initiates or inhibits environmental forces.

The right question to ask at this time is: ‘what is a model? Model is defined as a representation of real phenomenon (that may not be observed directly) using certain parameters (e.g. mathematical equations, physical laws, specialized alphabets, statistics, computational representation, simulation, animation etc.) to represent it (over a given period e.g. years, months, decades, century etc.) to make it easier to be comprehended. In actual fact, scientific model could be an image, simulation, diagram, or picture. It could be a physical model e.g. aircraft model. It could be visual representation in form of flowcharts; pictures that help improve the process of teaching and learning. It could be a computer program, or set of complex mathematics that describes a situation. Model is above all futuristic because it must be able to forecast or now-cast phenomenon or processes. The definition of environmental model is not significantly different from the basic definition of model. According to DBW [11], environmental modeling involves the application of multidisciplinary knowledge to explain, explore and predict the Earth's response to environmental change, both natural and human-induced. Environmental models are used to study climate, coastal changes, air pollution, aerosol loading, hydro-ecological systems, ocean circulation, surface and groundwater, terrestrial carbon-print, ecological systems etc.

1.1 Environmental Models in Existing Disciplines

In this section, the types of environmental model in some discipline shall be listed and explained. In the latter part of this section, the focus would be to discuss the conflict between environmental models as it cut across intra-discipline and inter-discipline. Most environmental models are presented in easy-to-use computer model that can be installed and used by novice. The challenge with the transformed environmental model (easy-to-use computer model) is the inability of the user to understand theoretical basics for developing the software. In other words, if the software is faulty, all the results obtained would be compromised with high errors. However, reputable institution or research center has certified some transformed environmental models in form of software. For example, the AERMOD can be regarded as an easy-to-use computer model that solves atmospheric dispersion model based on atmospheric boundary layer turbulence structure and scaling concepts. This software was developed by American Meteorological Society (AMS) and United States Environmental Protection Agency (EPA) in 1991. One of the uniqueness of AERMOD is its ability to handle flat or complex, rural or urban terrain and includes algorithms for building effects and plume penetration of inversions aloft. However, the basic principle of AERMOD relies on the Gaussian dispersion for stable atmospheric conditions and non-Gaussian dispersion for unstable conditions (high turbulence).

It should be noted that the list of model that is mentioned in this book were listed for illustration purpose only. Secondly, the type of environmental model mentioned in this chapter is captured under the main kind of environmental pollution, that is, air pollution, land pollution, and water pollution. Other pollution (visual pollution, light pollution, noise pollution, thermal pollution etc.) is acknowledged but not discussed. Thirdly, the environmental model that was mentioned was based on the basic needs of life forms. Some lists of environmental model (easy-to-use computer model) for water pollution include:

i. WQMCAL is a basic river water quality model that was prepared by Jolhnkai [12] in the framework of IHP-V Project of the United Nations Educational, Scientific and Cultural Organization. Theoretical formulation can be found in the reference [12];
ii. DESCAR is a software for wastewater dispersion analysis. The program calculates the pollutant concentration in each point of the water considering each one

of the pollutant sources and the conditions of the water. It is used for numeric simulations of water pollution processes;

iii. DISPRIN is a river water quality model software to simulate the impact of pollutant discharges on river systems. The model records for the primary scattering forms working in waterways and in addition the weakening from approaching tributaries and first-order kinetic decay processes. The model is dynamic and re-enacts the hourly conduct of waterway;

iv. Water Resources Water Quality Software (WRWQS) is an open-source software developed by the U.S. Geological Survey (USGS) to be used in the public interest and the advancement of science;

v. ESDV is a river self-purification modeller software for monitoring contamination levels in waterways. Its theoretical background is a numerical tool, developed based on the first order reaction of Biological Oxygen Demand (BOD) and the modified Streeter and Phelps equation, in order to determine the variations of Dissolved Oxygen (DO) and BOD along the studied river reaches;

vi. INCA is a stream quality software that is designed to simulate dynamics of biogeochemical and hydrological processes in stream systems. It is also used to assess a wide range of environmental change issues including land use change, climate change and changing pollutant loads in waterways;

vii. Like, the WRWQS mentioned in item four (i.e., above), SPARROW is a stream quality software developed by USGS. It works on the principle of non-linear, regression-based model for estimating and predicting pollutant concentration and transport, on basis of monitored concentration data and information on catchment characteristics;

viii. ODOURsim is a dynamic odour emission modeling software for wastewater treatment systems. It is commonly used for the modeling and simulation of odours generated in municipal and industrial wastewater treatment plants;

ix. H2OMAP SWMM is a powerful, stand-alone GIS-based computer program for hydrologic, hydraulic, and water quality simulation model for the effective management of urban storm-water and wastewater collection systems;

x. H2OMAP Water MSX is a software for modeling multiple interacting contaminants, that is, using water quality components rather than contaminants as well as sediment deposition and re-suspension in drinking water distribution systems.

Hydrology environmental models are developed models used to solve severe water challenges in urban and rural settlements. In this section few of the easy-to-use computer software for solving environmental challenges include:

i. Topmodel is stream hydrology software developed by Lancaster University for simulating overland flow, saturation infiltration, exfiltration, subsurface flow, evapotranspiration, and channel routing in a water basin;

ii. SHETRAN is stream hydrology and water quality software distributed freely by Newcastle University. It entails flexible, 3-D finite difference model, designed to simulate flow, and sediment and contaminant transfer in stream catchments;

iii. KISTERS software is professional water resource management software that provides groundwater managers with the data they need to effectively manage groundwater resources. Make better decisions by making sure you have timely, error-free and accessible data. The Hydstra version of KISTER is a non-relational database that supports SQL query;
iv. AGNPS is stream hydrology and water quality software developed by the US Department of Agriculture (USDA) to estimate pollution loads from agricultural watersheds; simulates surface water runoff, nutrients, sediments, chemical oxygen demand (COD), and pesticides from point and nonpoint sources of agricultural pollution;
v. AQUARIUS Time-series software is the most powerful platform for managing water resources. Environmental data from multiple sources are securely stored for fast, central access;
vi. XPSWMM is a dynamic hydraulic and hydrologic modeling software that combines 1D calculations for upstream to downstream flow with 2D overland flow calculations so that its users can see what truly happens to the storm-water system, foul water system or floodplain when waters flow, populations increase or catastrophic events hit;
vii. Modflow is groundwater hydrology and geochemistry software that is developed by USGS. It has the capability to simulate groundwater recharge, flow and solute transport;
viii. Mike-SHE is groundwater hydrology and water quality model developed by DHI technologies. It makes use of an advanced, integrated modeling system for simulating hydrological processes in linked surface and groundwater systems, including evapotranspiration, runoff, discharge, groundwater recharge and environmental fate of contaminants;
ix. WaterWare is water resources management information system software that is implemented in an open, object-oriented client-server architecture, fully web-enabled and Internet based, supporting the seamless integration of databases, GIS, simulation and optimization models, and analytical tools into a common, easy-to-use framework;
x. AQUASEA is software package developed to solve the shallow water flow and transport equations using the Galerkin finite element method. AQUASEA was first developed in 1983 to solve two-dimensional problems, and since 1992, it has been continuously upgraded and tested worldwide on the most difficult modeling problems.

One of the salient environmental model is the air pollution models. Like other models mentioned above, theoretical model has been transformed to easy-to-use computer software or models. Some of the air pollution model includes:

i. CERC—Version ADMS 5 is an industrial air pollution modeling Software that works on theoretical principles particulate-dispersion model. It is used for modeling air quality impact of existing and proposed industrial installations. Its unique features takes care of impacts of buildings, complex terrain, coastlines and variations in surface roughness, dry and wet deposition, and NOx chemistry schemes;

ii. THOR is a comprehensive air pollution forecast system, combining several air pollution models and one air forecast model. The system is used to produce operational 3-day air pollution forecasts, four times a day;

iii. CALPUFF is an advanced Lagrangian puff modeling system for the simulation of atmospheric pollution dispersion. It simulates the effects of time- and space-varying meteorological conditions on pollution transport, transformation, and removal;

iv. Air force dispersion assessment model (ADAM) is an easy-to-use computer model that works on the modified box and Gaussian dispersion model. Its unique feature incorporates thermodynamics, chemistry, heat transfer, aerosol loading, and dense gas effects;

v. HYSPLIT—National Oceanic and Atmospheric Administration (NOAA) and Australia's Bureau of Meteorology developed Hybrid Single Particle Lagrangian Integrated Trajectory Model. It is used for computing simple air parcel trajectories to complex dispersion and deposition simulations of atmospheric pollutants;

vi. Numerical atmospheric-dispersion modeling environment (NAME) is developed by the UK's Meteorology Office for local to global scale model. It is used for forecasting of air quality, air pollution dispersion, acid rain, tracking radioactive emissions, accidental air pollutant, and long-term environmental impact analysis. It is an integrated model that includes boundary layer dispersion modeling;

vii. The AERO-POLlution model was developed at the Tartu Observatory in Estonia. It uses Gaussian plume model for simulating the dispersion of continuous, buoyant plumes from stationary point, line and area sources over flat terrain on a local to regional scale, wet and/or dry deposition as well as the effects of buildings in the plume path;

viii. Electricite de France (EDF) developed MERCURE. The code is a version of the CFD software ESTET, developed by EDF's Laboratoire National d'Hydraulique. It is to perform air pollution dispersion modeling on local or urban scales. Its unique feature include Some of the models capabilities and features are point or line pollution point, continuous or intermittent pollution point, buoyant or dense gas plumes, and deposition or decay of plume pollutants;

ix. Environment Canada's Canadian Meteorological Centre (CMC) developed Trajectory software. It is to calculate the trajectory of a few air parcels moving in the 3D wind field of the atmosphere. Its theoretical background is rooted in advection transport theories;

x. RIMPUFF is a local-scale puff diffusion model developed by Risø DTU National Laboratory for Sustainable Energy, Denmark. It is used to deal with chemical, nuclear, biological and radiological releases to the atmosphere.

There are several environmental models for land pollution. Land pollution can be severe (as shown in Figs. 1.3, 1.4 and 1.5). It is gradually becoming a threat globally because it may lead to food shortage and the pollution of the aquifer (i.e. when pollutants are released to the ground and make their way down into groundwater). Land pollution may be caused by the following activities i.e. mining, erosion, oil spillage, forest fires, open burning of refuse, deforestation, concentration of population, land treatment without adequate evaluation etc. Also the reckless dumping of one or more of the following can cause land pollution: municipal solid waste, industrial waste, hazardous waste (chemicals or pet bottles), domestic waste etc. Figure 1.3 show land pollution that was caused by oil spillage. Most time, this challenge ravage oil producing communities. Udeze [13] itemize the oil spillage menace in the Niger-Delta region of Nigeria in an article titled “10 horrible effects of oil spill in Nigeria”. The risk are human health risk, damage to the ecosystem, health hazard to terrestrial and aquatic animal, risk to food security, loss of aquatic and terrestrial plants/vegetation, depletion of fish population, danger to wildlife, air pollution, acid rain and poverty. Figure 1.4 shows the land pollution that is caused by refuse dumping. This practice is common in rural and unorganized urban setting. Figure 1.5 shows the land pollution that is caused by industrial waste. Like oil spill, the impact of the land pollution extends beyond the physical representation shown in Fig. 1.4.

Many environmental models have been proposed [14, 15]. Some of the models have been transformed to easy-to-use computer model or software. Few of this software are discussed below.

Fig. 1.3 Land pollution—caused by oil spillage

Fig. 1.4 Land pollution—caused by refuse dump

Fig. 1.5 Land pollution—caused by industrial waste

i. 3DFATMIC is unix based software developed by US Environmental Protection Agency, Office of Research and Development, National Risk Management Research Laboratory, and Center for Subsurface Modeling Support. It is used to simulate subsurface flow, transport, and fate of contaminants that are undergoing chemical and/or biological transformations. Theoretical inclination of the model is the Galerkin finite element solution of Richard's equation. The transport module is a hybrid Lagrangian-Eulerian approach with an adapted zooming and peak-capturing algorithm.

ii. Soil Vision Pro is soil pollution monitoring software for monitoring and testing. It has a database of over 6200 soil samples. 98% of these soil samples have a soil-water characteristic curve measured in a laboratory. The soil database also contains saturated permeability (hydraulic conductivity) data on over 2500 soils as well as unsaturated permeability data on over 700 soils.

iii. Kartotrak is contamination characterization software for contaminated sites characterization. It can also be used to locate and estimate contaminated soil volumes.

iv. CalTOX is contamination control software. California Department of Toxic Substances Control develops it to estimate the concentration of a chemical in soil to the risk of an adverse health effect for a person living or working on or near the contaminated soil.

v. CHEMFLO is contamination transport software that is built on the DOS environment. The US Environmental Protection Agency, Office of Research and Development, National Risk Management Research Laboratory, and Centre develop it for Subsurface Modeling Support. Like 3DFATMIC software, CHEMFLO operates on theoretical background of the Richards equation (water) and the convection-dispersion equation (chemicals). It is used to simulate water movement and chemical transport in unsaturated soils.

vi. Emsoft is exposure model software for soil-organic fate and transport. It is developed by US Environmental Protection Agency and National Center for Environmental Assessment (NCEA). The software is used to: determine concentrations of contaminants remaining in the soil over a given time; quantify the mass flux (rate of transfer) of contaminants into the atmosphere over time; and subsequently calculate contaminant air concentrations by inputting mass flux values into atmospheric dispersion models.

vii. UnSat Suite Plus is groundwater contamination software that can be used in land pollution sites. It used to determine groundwater contamination problems at either surface spills or leaks from surface or near surface storage tanks and facility operations.

viii. WinEPIC is erosion-productivity impact calculator software developed by US Department of Agriculture and Natural Resources Conservation Service. It is used to assess the effect of soil erosion on soil productivity.

ix. MMSOILS is DOS environment software for estimating the human exposure and health risk associated with the releases of contaminants from hazardous waste sites. It was developed by the US Environmental Protection Agency and Center for Exposure Assessment Modeling (CEAM). It is used to estimate the transport of a chemical in groundwater, surface water and soil erosion.

x. NAPL Simulator operates on the DOS environment. It is a software that estimates the Non-Aqueous Phase Liquids (NAPLS). The US Environmental Protection Agency, Office of Research and Development, National Risk Management Research Laboratory, and Center developed it for Subsurface Modeling

Support. NAPL is used for simulation of the contamination of soils and aquifers that results from the release of organic liquids.

The next session of interest is to discuss the green areas in environmental modeling in some selected disciplines. This part of the work is to raise the curiosity of the reader to start developing easy-to-use computer software or model. Basically, the transformation from concept-development to the final product i.e. easy-to-use computer software or model is demanding. By experience, it is easy to sub-divide the process into eight, that is,

i. Concept-development is the beginning of the work. The modeller at this stage is trying to choose a research topic based on glaring statement of problem.
ii. Model-formulation is the second part of the modeling process. The modeller chooses the mathematical or statistical or computation tool to use in order to accomplish the set-tasks. From experience, most beginners in environmental modeling get stock because of inadequate technical skills. The immediate solution may be to seek counsel from experienced colleagues.
iii. Data-generation is the third process. At this stage, the modeller gathers reliable dataset to evaluate aspect or whole of the modeller's hypothesis.
iv. Model-validation is the fourth step of the process. This aspect is very important to help the modeller gain significant level of confidence in further analysis. Experience modeller at this stage seeks for large dataset that was generated at various conditions to test the model.
v. Model-re-evaluation is the fifth step of the process. The modeller attempts publishing or presenting the model to arouse the interest of researchers. This process may be very complicated because main aspect of the model is not expected to be divulged to avoid hijacking of the modeller's work. Also, some modeller involves the assistance of established research center with capable hands to expand on the findings. This route may be the best but may be confiscated if there are no legal professionals to guide the process.
vi. Choice of computer license is the sixth step of the process. Here, the modeller must be certain if the computer model must be guided by the open-source license or the closed/purchased source license. Most modeller chooses the open-source license if the framework of the computer model or software would require external libraries that was not built by the modeller. The duration of this process depends on the financial capability and technical know-how of the modeller.
vii. Software-evaluation is the seventh step of the process. The modeller tests the computer model or software to affirm its accuracy with previous simulation. The modeller is expected to document the challenges before releasing to the first set of users.
viii. Comparative-analysis is the last step of the process. The modeller compares his/her computer model or software with existing ones. The modeller is expected to document the observations to guide additional features for the new computer model or software.

Most institution lump the civil engineering and environmental engineering discipline. Two decades ago, environmental engineering was a sub-division of civil engineering. Environmental engineering is primarily based on societal sustainability via the use of water, land and air resources. This goal is achieved by managing these resources so that environmental pollution and degradation is minimized. Hence, environmental engineers study water, soil and air pollution problems, and develop technical solutions needed to solve, attenuate or control these problems in a manner that is compatible with legislative, economic, social and political concerns [16]. On the other hand, civil engineers are particularly involved in activities such as water supply and sewerage, management of surface water and groundwater quality, remediation of contaminated sites and solid waste management. Some of the environmental models associated to civil/environmental engineering are in the specialized area of construction waste disposal [17], environmental impact assessment, pollution study of river and channels, recycling industrial effluent, oil spill on marine environment, solving the e-waste problem, aqueous phase geochemical characterization, analysis of bioclimatic building design, long-time hydrological systems etc.

Construction waste management (CWM) has received worldwide attention in the European Union member countries, Hong Kong and United Kingdom [17]. CWM primarily looks at the best way to utilize building waste like rock and soil, waste asphalt, bricks, concrete, plasterboard, timber and vegetation, asbestos, bamboo wood, shreds from metallic roof and contaminated soil [17, 18]. In most developing countries, construction has contributed to the number of deaths due to the release of dangerous chemicals into the air or percolates through the ground to aquifer. Environmental impact assessment is important in estimating future challenges as it is inter-disciplinary in nature and operation [19–21].

Environmental biologists think about particular territorial conditions and the living beings and untamed life that occupy those areas. These experts help in the assurance and survival of untamed life inside any given biological system, and they evaluate the influence of human activities on the environment. In environmental biology, the current modeling topics include:

Marine community ecology, population and community ecology, bio-assessment of freshwaters, tropical forest ecology, biological invasions, soil sciences, insect-plant Interactions, ecological modeling and QGIS, tropical wetland ecology, food borne pathogens, wildlife ecology and management, peat lands and environmental Change, environmental pollution etc.

Environmental chemistry is the logical investigation of the chemical and biochemical phenomena that happen in common places. It is an interdisciplinary science that incorporates atmospheric, oceanic and soil science, and how they vigorously depend on expository science of the environment. On the other hand, Nature [22] defines environmental chemistry is the study of chemical processes that occur in water, air, terrestrial and living environments, and the effects of human activity on them. It includes topics such as astrochemistry, atmospheric chemistry, environmental modeling, geochemistry, marine chemistry and pollution remediation. Currently, emerging models in environmental chemistry are: systemic exposure of human population to chemicals; bio-monitoring of organic contaminants on coastline communities;

human exposure to toxic mixtures; persistent organic pollutants in food; aerosol pollution in industry; biochar growth and bioavailability of conazole fungicides in soil; chemical toxicity of rubber waste on marine ecology; chemicals and their health risks in human population; chemical exposure on human metabolism; microfluidics—lab on a chip in environmental research; nutritional and intestinal microflora and their impacts on the human health; molecular modeling of enzymes' substrate specificity; nanopesticides in the environment; hermetia illucens to waste processing etc.

The field of environmental and atmospheric physics are almost synonymous. Some institutions group the two courses together. The main difference is that atmospheric physicists seek to model Earth's atmosphere and the atmospheres of the other planets using fluid flow equations, chemical models, radiation budget, and energy transfer processes in the atmosphere (as well as how these tie into other systems such as the oceans) while the environmental physicist attempt model the measurement and analysis of interactions between organisms and their environment. The current trend of environmental model in environmental/atmospheric physics includes: optical and chemical properties of mineral dust and carbonaceous dust aerosols; photochemical dispersion models; analysis of FTIR and UV-vis spectrometers; gas phase and heterogeneous chemical processes and reaction dynamics; aerosol generation & characterization instrumentation; measurement of the mass concentration of aerosols; dispersion model; irradiative transfer models; measurement of the vertical distribution of atmospheric ozone; Mesoscale Meteorological models (PSU/NCAR—MM5 and WRF); Oceanic impact on mid latitude climate variability; global climate model ECHAM5/Messyv1.4—general circulation model; Coupling of oceanic and atmospheric heat transport; carbon cycle.

Meteorology is a branch of the atmospheric sciences that focuses on weather processes and forecasting. It includes discipline as atmospheric chemistry and atmospheric physics. Meteorologist use physics and chemistry principles to analyze, evaluate, forecast, now-cast, investigate, examine etc. According to World Meteorological Congress (WMO), the focus of meteorologist should be on the highlighted priorities for weather and climate research [23]. The five priorities are: deliver science for services; build seamless (predictive) models; improve infrastructure (by providing computing infrastructures and facilities); nurture a diverse workforce; and share ideas (among existing and new network of researcher). Existing models in meteorology are mainly in the following areas: air pollution/urban meteorology; climate change—atmospheric dynamics; fire weather and wildfire dynamics; mesoscale processes; mountain meteorology; tropical meteorology; weather systems and forecasting; wind energy assessment; storms, floods and droughts assessment. Current research trends in meteorology that can be modelled includes: formation, structure, and evolution of EF4 tornado; influence of the northeast cold vortex on flooding; cycle of rainy season at different types of ENSO; atmospheric circulation patterns over regions or continent; diurnal and seasonal variations of thermal stratification and vertical mixing in lakes and rivers; Impact of Tropospheric Ozone on climate systems; chemical components, variation, and source identification of particulate matter; tracking severe pollution event in mega-cities; estimation of crop evapotranspiration using satellite remote sensing-based vegetation index; spatial downscaling of satellite precipitation measurement etc.

Environmental management is concerned with physical, social and monetary condition of venture or undertakings. It looks to create incorporated frameworks for environmentalist. Environmental management can be divided into sub-division like environment and enterprise objectives, scope and structure of the environment, interaction of nature, society and the enterprise, environmental impact assessment, economics of pollution, prevention, environmental management standards, environmental audits etc. Current research trends in environmental management are: evaluation of solid waste management; supporting environmental good practice on site through e-learning; investigations into effects of land-use change on decline in biodiversity and species extinctions; research into the long-term behaviour of land use impacts on diffused water pollution; sustainable water uses in developing countries; cost involvement on desalinisation in developing-world context; working with wildlife—supporting effective management on construction projects; financial insight in reducing pollution wildlife parks etc.

1.2 Special Case Study: Aerosol Models

It is quite impossible to discuss environmental model in general terms because it is a very broad subject that is still expanding in scope as new disciplines emerge. The most prominent aerosol model is the atmospheric aerosol model. Atmospheric aerosol particles include various chemical species, such as sulphate, black carbon, organic carbon, mineral dust and sea salt. The atmospheric aerosol model is the most complicated model in environmental studies because:

i. It requires mathematical skills in the formulation. The mathematical challenges originate from its open system. Recall in physical laws, a closed system is a system whose input and output parameter is known. It is easier to use mathematical equations relating to conservation of mass or energy to discuss particulates distribution and dynamics. Atmospheric aerosol model is in an open source system and controlled horizontally by wind forces and vertically by buoyancy forces. The modeller is most faced with the reality of dead-end discussion because of many events and activities that was not factored in the model formulation.
ii. It is affected by climate change. Hence, it is possible to lose the sensitivity of the model at short duration. It is based on the above many new atmospheric aerosol has emerged in scientific community.
iii. It requires a very large dataset to understand the regional trends of atmospheric aerosol deposition, loading and retention. When this trends are understood, the modeller starts seeking for answers why these trends are sustained or dispel at particular periods in the year. This process is quite important because it does

not only require mathematical skills. It requires adequate skill in meteorology. Unfortunately, meteorological models are dynamic too.

iv. The classification of atmospheric aerosols with respect to type, size, lifetime and transport in model formulation is quite complex. At the moment, over 76% of the world's communities do not have access to ground measuring equipment. Hence, most scientists adopt satellite remote measurements. The satellite measurement dataset may differ from the ground measurement sometime [10].

v. It is constantly subjected to different atmospheric forces that change the orientation of the atmospheric aerosols along altitudinal profiles. Hence, adequate understanding of the atmospheric vertical profile must be known. Some modeller threads the safe zone by only defining research scope (i.e. a specific height in the troposphere).

It is important to understand what atmospheric aerosols is all about, before going into details about its metamorphosis. Atmospheric aerosols (or particulate matter) are solid or liquid particles or both (with diameters between about 0.001 and 120 μm) suspended in the atmosphere. Aerosols interact both directly and indirectly with the Earth's radiation budget and climate because of its high varying size, source, chemical composition, lifetime, amount and distribution in space and time. The main sources of atmospheric aerosols are volcanic aerosols, desert dust, and human-made aerosol. These atmospheric aerosols affect the earth's climate in significant patterns. For example, volcanic aerosol layer is majorly made-up of sulphur dioxide gas and it deposits at the stratosphere and is distributed by wind forces to distant geographical enclave (see Fig. 1.6).

The second pattern of aerosol emission that affects global climate is streaming out of dust veils over the Atlantic Ocean from the deserts of North Africa. The distribution of desert dust extends to various locations on the American and Asian continent (see Fig. 1.7). The third pattern aerosol affects global climate is through human-made aerosols from anthropogenic activities. A large fraction of human-made

Fig. 1.6 Global volcanic aerosol distribution [24]

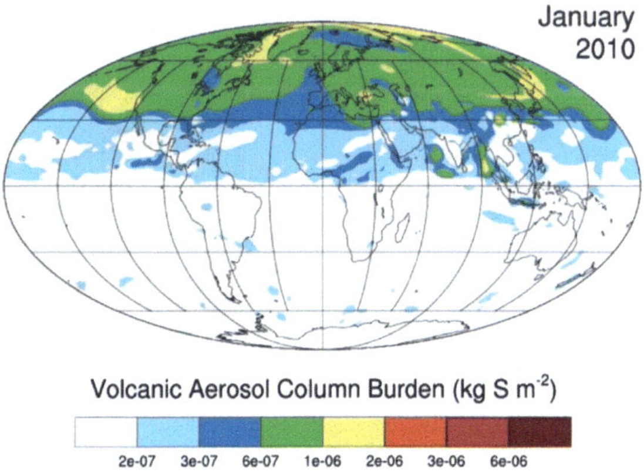

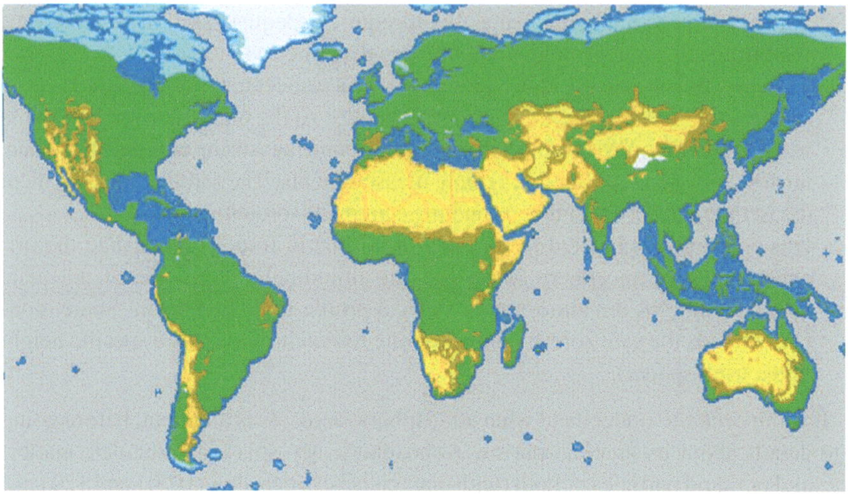

Fig. 1.7 Global desert dust distribution [25]

aerosols come in the form of smoke from burning tropical forests, gas flaring and burning of coal and oil (see carbon dioxide distribution in Fig. 1.8).

Parameters that should be in a good aerosol model include: emissions, atmospheric transport, chemical reactions, gravitational settling, removal processes by dry deposition and precipitations. Some aerosol models are described below:

i. MRI-CGCM3 is a new global climate model of the Meteorological Research Institute, Tsukuba, Japan [27]. MRI-CGCM3 is composed of atmosphere-land, aerosol, and ocean-ice models, and is a subset of the MRI's earth system model MRI-ESM1.
ii. MRI-CCM2 is an atmospheric (Ozone) chemistry model developed by Meteorological Research Institute, Tsukuba, Japan. It is used for predicting global

Fig. 1.8 Global carbon dioxide distribution [26]

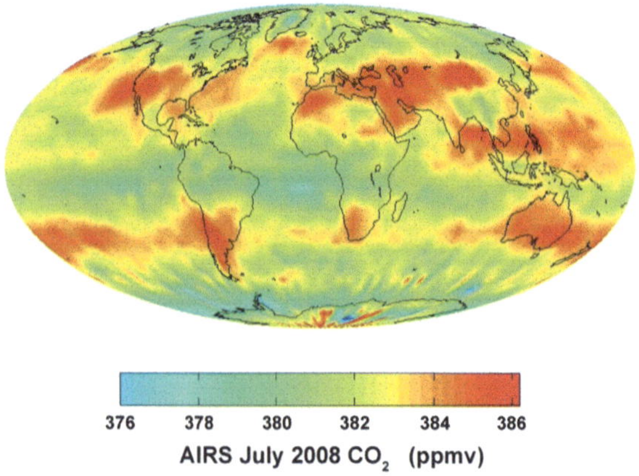

distributions of atmospheric trace gases such as ozone along with chemistry–climate interactions. It is used to simulate the following processes: chemical conversion of trace gases, (grid scale) advective transport, (sub grid scale) convective transport and boundary layer diffusion, dry and wet deposition, and emissions.

iii. MASINGAR mk-2 is software for aerosol chemical transport model. It is used to determine the chemical transport of five type of aerosols i.e. sulphate, black carbon, organic carbon, mineral dust and sea salt.

iv. REMO-HAM is a regional aerosol-climate model that is developed by University of Eastern Finland. It is based on the REMO regional climate model and includes most of the major aerosol processes. The structure for aerosol is similar to the global aerosol-climate model ECHAM5-HAM.

v. CMAQ-UCD Aerosol Model was developed by University of California at Davis (UCD). It is used to simulate air quality in the coastal urban location through the adoption of SAPRC99 gas-phase chemical mechanism. This is possible because it use sectional aerosol module with 12 species in each size bin, including SO_4, NH_4, NO_3, Na, Cl, elemental carbon, primary organic aerosol, biogenic secondary organic aerosol (SOA), anthropogenic SOA, mineral dust, water, and hydrogen ion.

vi. PDRMIP is a new climate model intercomparison initiative and was launched in Oslo in November 2013. In PDRMIP a number of different climate models will be used to explore whether differences in precipitation at present and future projections can be linked to differences in forcing mechanisms.

vii. ECHAM-HAM is a global aerosol-climate model. It was developed at the Max Planck Institute for Meteorology [28]. It is used to: simulate tropospheric and stratosphere aerosols; calculate irradiative effects of aerosols on the atmospheric dynamics; simulate aerosol optical depth (AOD) and extinction profiles; estimate aerosol absorption featured negative biases; calculate aerosol nucleation and water uptake; perform an explicit treatment of secondary organic aerosols; analyse modified emission calculations for sea salt and mineral dust; estimate the coupling of aerosol microphysics to a two-moment stratiform cloud microphysics scheme; perform alternative wet scavenging parameterizations.

viii. AeroCom (Aerosol Comparisons between Observations and Models) is a model that fosters the advancement of the understanding of the global aerosol and its impact on climate. It is comprised of a large number of observations and results from more than 20 global models. The model or project is anchored by World Climate Research Programme (WCRP) with the collaboration of: The Working Group on Coupled Modeling (WGCM); The Working Group on Subseasonal to Interdecadal Prediction (WGSIP); The Working Group on Numerical Experimentation (WGNE); and World Weather Research Programme (WWRP).

ix. EUCAARI (European Integrated project on Aerosol Cloud Climate and Air Quality interactions) is a EU Research Framework 6 integrated project focusing on understanding the interactions of climate and air pollution. The project has the participation of 48 partners from 25 countries. The project was used to: distinguish and measure the procedures and sources influencing global and regional aerosol concentrations; measure the physico-compound properties of atmospheric aerosols; measure the feedbacks that connects aerosol loading to climate change and atmospheric aerosol concentration.

x. ATRAS2 (Aerosol Two-dimensional bin module for foRmation and Aging Simulation version 2) works on the aerosol box model. It is one the emerging global aerosol model. It is used for: simplifying the treatments of aerosols processes e.g. condensation of sulphate, ammonium, and nitrate, organic aerosol formation, coagulation, and new particle formation processes; analysing treatment of chemical compositions using two interactive bin representations; calculating number concentrations, size distributions, and BC mixing states and their gradual changes by aging processes [29].

xi. TCAM (Transport Chemical Aerosol Model) is a multiphase three-dimensional Eulerian grid model designed for modeling dispersion of pollutants (in particular photochemical and aerosol) at mesoscale. Environmental Systems Modeling and Assessment group of University of Brescia develop it. Its principle of the TCAM is based on the mass-balance, transport-diffusion, emissions, deposition and advection module [30].

xii. CARMA (Community Aerosol and Radiation Model for Atmospheres) is a general-purpose sectional microphysics code that has been used to study a wide variety of aerosols in planetary atmospheres. It was developed by Turco et al. [31] and Toon et al. [32]. Also, the model is able solve gas phase sulphur chemistry, aerosol microphysics and three dimensional aerosol advection. Also, CARMA is used to estimate or analyse cloud and aerosol on Earth, as well as those on Venus, Mars, Titan, and exoplanets. The standard versions of the model are maintained at National Centre for Atmospheric Research (NCAR) and distributed to wide community of users.

xiii. DHARMA (Distributed Hydrodynamic Aerosol and Radiative Modeling Application) is a large-eddy simulation code. DHARMA was developed by David Stevens (Stevens and Bretherton, JCP, 1996). Its application includes aerosol and cloud microphysics and radiative transfer components that are based on CARMA, which required making the microphysics model suitable for use in a 3D framework. The radiative transfer problems are solved using the two—stream method [32]. It is also used to solve inelastic treatment using fully compressible equation system.

xiv. WACCM3 (Whole-Atmospheric Community Climate Model) is a comprehensive numerical model, spanning the range of altitude from the Earth's surface to thermosphere. It is developed by an inter-divisional collaboration that unifies certain aspects of the upper atmospheric modeling of High Altitude Observatory (HAO), the middle atmosphere modeling of Atmospheric Chemistry Observation and Modeling (ACOM), and the tropospheric modeling of Climate and Global Dynamics (CGD), using the NCAR Community Earth System Model (CESM) as a common numerical framework. The chemical mechanism in WACCM includes comprehensive tropospheric, stratospheric and mesospheric chemistry.

xv. GLOMAP is a flexible aerosol microphysics model. It was developed in collaboration of Institute for Climate and Atmospheric Science (ICAS). It works on two principles i.e. GLOMAP-bin (which controls the comprehensive sectional (bin) model) and GLOMAP-mode (that is a faster modal version). It is used to: calculate the details of the aerosol number size distribution on a global scale; initiate faster code for inclusion in climate and weather models. The GLOMAP aerosol model is used in several models for climate, earth system, weather, and air quality applications. GLOMAP is in different host models e.g. TOMCAT Global Chemical Transport Model, HadGEM-UKCA and European Centre for Medium-Range Weather Forecasts Integrated Forecasting System. The TOMCAT CTM is a global chemistry and aerosol model that uses analysed meteorology techniques to transport and remove the aerosol without including feedbacks on the weather and climate. UKCA is a chemistry-aerosol-climate model built on the Met Office Unified Model (UM). ECMWF-IFS are the comprehensive earth-system model developed at ECMWF solves data assimilation and forecasting activities [33].

xvi. NAAPS (Navy Aerosol Analysis and Prediction System) is a global aerosol model developed by Naval Research Laboratory (NRL). It is used for: operational dynamics, e.g. 'REAL' weather; 120-hour forecasts; operating in near-real-time; global coverage; dust simulations; smoke simulations; improving the dust source function; verifying the sulphate simulations; improving the microphysics and chemistry.

xvii. RAMS (Regional Atmospheric Modeling System) is a code (written almost exclusively in FORTRAN 77) for simulating and forecasting meteorological phenomena, and for depicting the results. Colorado State University (CSU) develops it. Its principles revolves around primitive dynamical equations which govern atmospheric motions, and supplements these equations with optional parameterizations for turbulent diffusion, solar and terrestrial radiation, moist processes including the formation and interaction of clouds and precipitating liquid and ice hydrometeors, sensible and latent heat exchange between the atmosphere, multiple soil layers, a vegetation canopy, surface water, the kinematic effects of terrain, and cumulus convection. It is used to: perform actual atmospheric simulations; predict the mixing ratio and number concentration of cloud droplets, drizzle, rain, pristine ice, snow, aggregates, graupel, and hail; analyze package which prepares initial data for the atmospheric model from observed meteorological data; post-processing model visualization and analysis package which interfaces atmospheric model output with a variety of visualization software utilities [34].

The aerosol models that are listed above are not the only aerosol models available. It should be noted that the models were arranged in no particular order for illustrational purposes.

1.3 Emerging Aerosol Models

The emerging models were mostly developed because of four main reasons, that is: correcting the flaw of an old model; improving on the functionality or features of an existing; developing new concepts due to new statement of problem; and writing to contribute to the body of knowledge. In recent times, emerging models are released based on improving on the functionality or features of an existing model. For example, most of the aerosol models mentioned in Sect. 1.2 are improved version of an existing model. In this section, the interest is reporting new model that exist due to the imperfection that is noticed in existing models and outdated version of models due to current issues.

So what are the new bugging issues in aerosol modeling? Saleeby and van den Heever [35] identified new issues in the Regional Atmospheric Modeling System (RAMS) that borders on the inability of old RAMS aerosol modules to perform: multiple aerosol modes, variable solubility, nucleation scavenging, regeneration, scavenging by dry and wet deposition, aerosol–radiation interactions, and/or the ability to track the temporal and spatial variability in activated, scavenged, and in situ aerosols by type. The improved version of CSU-RAMS was proposed in Saleeby and van den Heever [35]. The new CSU-RAMS works on nine aerosol species: sub micrometer and super micrometer modes of ammonium sulphate, mineral dust, and regenerated aerosols and three modes of sea salt. It has shown tremendous insight for further research exploits [36].

The science that relates atmospheric aerosols and cloud formation are still unclear based on very many events that occur at different climatic systems. There arose the need to have a reference thermodynamic model for gas/liquid/solid equilibrium calculations. Based on the above, the Extended Aerosol Inorganics Model Project (E-AIM) was established to address this challenge. E-AIM is a compendium of different models. For example, it has phase equilibrium models of aerosol systems made available on this web site can be used to calculate the equilibrium state of chemical systems containing mixtures of organic compounds, water, and the ions H^+, NH_4^+, Na^+, SO_4^{2-}, NO_3^-, Cl^-, and Br^-. Also it has models that enable the distribution of water and the other components to be calculated between liquid, solid and vapour phases for ambient conditions (temperature, relative humidity) specified by the user. E-AIM models were developed with support from the Natural Environment Research Council (NERC). Till date, there are still improvements made to the E-AIM models. For example, Models II to IV of the E-AIM models can now be used to calculate "Köhler curves" which represent the equilibrium saturation (or supersaturation) of water over liquid droplets of known size.

The importance of thermodynamics of atmospheric aerosols is been emphasized by renowned scientists, especially, on the need of a more robust model that will computationally predict the physical state and composition of inorganic atmospheric aerosol. One of the early responses to this emerging need was the ISORROPIA. This model is an emerging thermodynamic equilibrium model for multiphase multicomponent inorganic aerosols. At the moment, there is the ISORROPIA and ISORROPIA

II that were developed by Division of Marine and Atmospheric Chemistry (DMAC) and Georgia Institute of Technology respectively. The models are used to examine the behaviour of four types of tropospheric aerosol (marine, urban, remote continental and non-urban continental). Also, the models can be used to calculate the composition and phase state of an ammonia -sulphate -nitrate -chloride -sodium -calcium -potassium -magnesium -water inorganic aerosol in thermodynamic equilibrium with gas phase precursors.

In the light of the above, there arose the need to have an elaborate view on global sectional multi-component aerosol model that can shed more light on carbonaceous aerosols that majorly occur via bush burning, coal burning, anthropogenic emission etc. One of the emerging models that seek to address this issue is the organic and inorganic spectral aerosol module ORISAM-TM4. Orisam-TM4 (like ORISAM-RAD) was derived from the ORISAM module which was developed by Bessagnet and Rosset [37] at the laboratory of Aerologie and global chemistry-transport model (CTM) TM4 [38]. ORISAM was originally designed to: for solving aerosol multicomponent systems, describe or estimate microphysical processes, and analyze processes of nucleation, coagulation, condensation, adsorption, chemical reaction and deposition. ORISAM-TM4 is an emerging global sectional multi-component aerosol for working on aerosol size distributions with a variable number of diameter sections (between 0.04 μm and over 10 μm) and detailed gas schemes on organic/inorganic chemistry aerosols. Guillaume et al. [39] used the ORISAM-TM4 to work on secondary organic aerosol (SOA)—carbonaceous BC (black carbon) and OC (organic carbon) aerosols. With the incorporation of the SOA calculations into ORISAM-TM4, the model can now estimate organic carbon without using the proportionality estimation of primary organic carbon. However, details on the size distributions are unclear. Scientists went further to seek a better way of addressing the limitations i.e. detailed description of time and spatial evolution of atmospheric particulate matter. SIREAM (size resolved aerosol model) emerged to give insightful contribution by solving the General Dynamic Equation for aerosols that relates to heterogeneous reactions at the aerosol surface, mass transfer between aerosol and cloud droplets, aqueous-phase chemistry inside cloud droplets, condensation/evaporation, coagulation, nucleation, inorganic and organic thermodynamics [40].

There was the need to focus on the multi-component aerosol over West Africa, especially, based on the high aerosol optical depth that can be seen via satellite imagery. Unlike other developed regions of the world, 90% of the communities do not have access to aerosol ground dataset. The only access to understand the physical properties of atmospheric aerosol in West Africa is via satellite remote sensing. Unfortunately, most satellite dataset along the coastal regions of West Africa are massively plagued with huge data loss. The West African regional scale dispersion model (WASDM) emerged to synchronize ground and satellite measurements in known locations. This finding led to the mathematical modeling of a new dispersion model that was able to calculate the aerosol loading over fifty communities in West Africa [1]. Other parameters that can be obtained from the model include aerosol retrieval factor, atmospheric constants, aerosol deposition and aerosol size distribution.

1.4 Big Data in Environmental Modeling

Big data is informational dataset that are so huge and complex, such that ordinary dataset-processing application (e.g. Excel) is insufficient to manage them. It is a term that portrays the vast volume of information—both organized and unstructured. Be that as it may, it is not the measure of information that is vital but what organizations, institutions or associations do with the information. In environmental studies, big data is common to some specific areas because of its long existence or sources of data generation. The environmental public health sector has huge data arising from toxicological issues from industrial pollution and medical issues from environmental issues. Geographic data is almost the most massive due to large database like geographical information system, satellite imagery and aerial imagery. Atmospheric monitoring data is also huge because of the variety of parameters that are considered. The sources of atmospheric monitoring dataset are satellite and ground measurements that have existed for more than four decades. The main advantage of big data in environmental studies is to mitigate errors, improve data-processing refinements, aid decision-making, promote accuracy, and manage forecast and now-cast events or activities.

When discussing big data, it is important to note the use of five terms and what they mean. The five terms are volume, velocity, variety, variability and complexity [41]. Volume refers to gathering information from an assortment of sources, including business exchanges, internet based life and data from sensor or machine-to-machine information. For example, after designing your experiment, the sources of dataset and what type of parameters that should be in the data makes-up the volume of the big data. Velocity in Big data refers to rate at which the dataset is generated within a pre-determined process. This parameter helps the industry, institution or researcher to plan adequately on how to analyze the dataset as soon as they are produced. For example, if an experiment generates 1 TB in sixteen hours and the process is expected to run for sixteen days without stop, the analysing tool should be able to process the information within the period of generation. The velocity of 'Big data' is quite important in choosing the investigation, probe or analysis tools.

The third parameter that describes 'Big data' is variety. Variety refers to how information comes in a wide range of arrangements—from organized, numeric information in traditional databases to unstructured content reports, email, video, sound, stock ticker information and budgetary exchanges. Variability in Big data refers to the changing nature of velocity and volume of dataset. For example, 1 TB may not generate every sixteen hours under certain conditions i.e. the volume can increase (say to 1.2 TB) or decrease (say to 0.65 TB). So if the variability of big data is known, it is easier to plan for recovery operational time when working on a continuous process that generates huge dataset. The last parameter of big data is the complexity. Complexity in Big data refers to the difficulty of associating the big data from two or more sources. Therefore, it is important to interface and correspond connections, chains of command and different information linkages or such information can go out of control. Planning a research work on big data is a conscious activity that requires the five parameters to be settled before choosing data source, investigation tool and mode of reportage. The technical issues on Big data are: data storage facilities, methods of capturing and transfer of data, choice of investigative tool, inadequate knowledge of data generation (secondary users of data), visualization and querying of data, data security, synchronizing similar dataset from different sources, interpretation of data, transparency or reproducibility of dataset, and ability of researcher to develop indigenous investigative tools.

In this section, the focus is to make the reader to understand the new trends of big data. This would enable the reader to understand how interesting or complicated research is becoming in recent times. The Big data cases that were discussed in this section were selected for illustration only. There may be better big data cases in environmental studies that was not discussed because of the authors proximity its resource or limited space to discuss each section.

Ponce Romero et al. [42] worked on the adoption of new big data tools and computational technologies to the water utility sector in addressing specific challenge of Wastewater Treatment Plants (WWTP). The researcher selected sources of the Big data i.e. residential, commercial and industrial points of origin, agricultural and hard surface runoff. This source defines the volume of big data that was generated at the end of the study. The parameters that were considered within each data source are pollution that originates from herbicides, pesticides, phosphorous, pharmaceuticals, and a new generation of chemicals of emerging concern. The complexity of the big data that was considered was contaminants arriving at the WWTP. Like it was discussed earlier, complexity can lead to variability. Environmental scientist, designer or modeller needs to understand this aspect to avoid over burdening of the whole exercise. In the case currently considered, Ponce Romero et al. [42] used the variability of the Big data to support evidence-based decision making and investment planning. The challenges associated with the applicability of existing computational analytical approaches were considered before choosing the investigating tools. In the case of the authors, the investigative tools (CIP and SAGIS) were developed. Figure 1.9 helps reader to appreciate the modeller mind-set when developing the investigative tool. Chemicals Investigation Programme (CIP) was used to monitor chemicals most likely to reach water treatment plants and the characterization of final effluents of 162 WWTPs from England, Scotland, and Wales. Source Apportionment-GI (SAGIS) outputs were used to predict determinants, contributions to modelled water concentrations and comparison of simulated outputs with observed river monitoring data. The authors were able to show that big data approaches are effective in the merging of diverse and large datasets to provide accurate and precise outputs assist that will assist audience in the water sector.

There are cases in big data research that the researcher does not have control over the volume, variety, variability, velocity and complexity of the dataset. For example, satellite measurements have been for almost four decades and the huge data that exist have no input from the secondary users. Hence, secondary users constrained to either develop the investigative tool or work with existing tools. The danger of using existing tools without knowing the '4 V's' (i.e. volume, variety, variability and velocity) used by the developer may not give accurate results. In recent times, renowned information

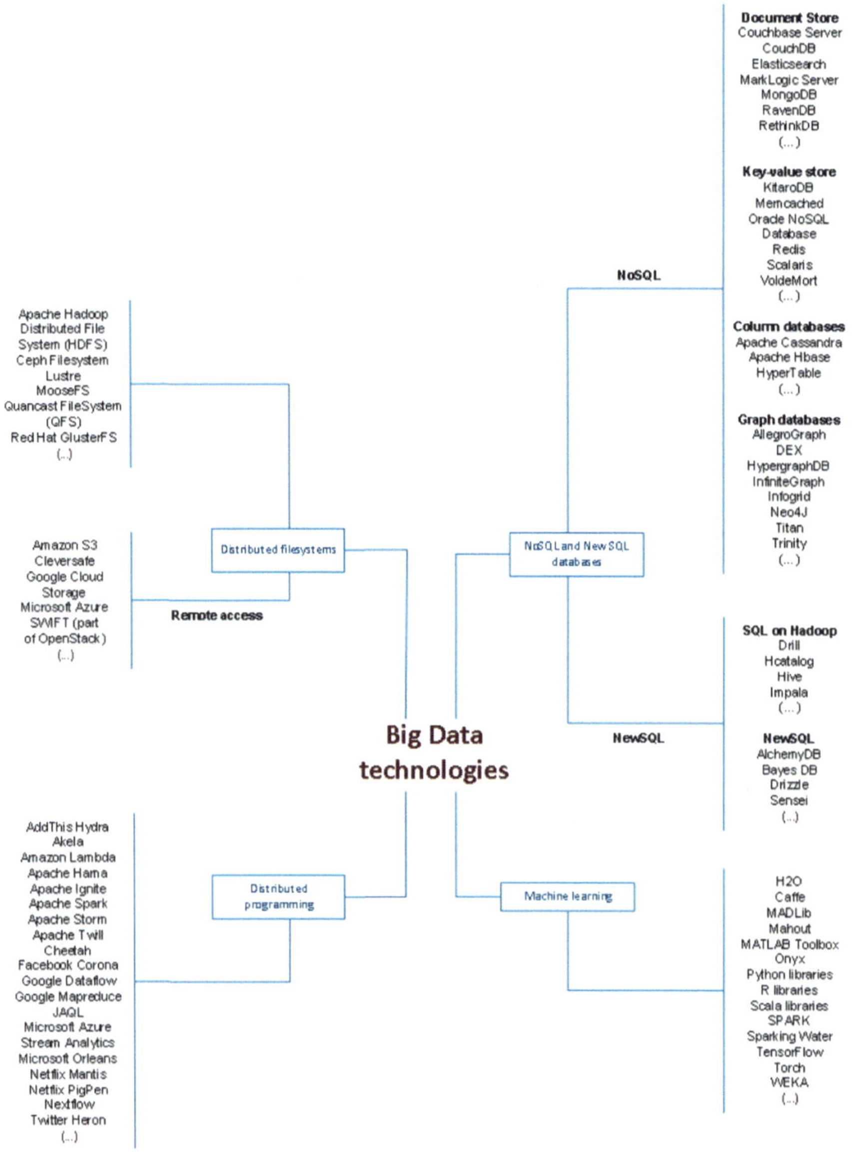

Fig. 1.9 Investigative software in big data analysis [42]

technology (IT) companies have come-up with the core platform where scientists can build up their investigative tool.

For example, A PaaS platform like IBM Bluemix enables user rapid development and deployment capabilities. Its architecture allows user to build resilient applications on an infrastructure with processing capabilities that scale automatically as needed [43]. In this book, one of the objectives is to help readers develop their own investigative tool. Users can progress using whatever platform that is easier for them. What should I do if I am a secondary user to a dataset? The cue to this question will be illustrated by an existing research in that line.

Varotsos and Krapivin [44] provided a description of the approach and methodology to be used to solve the problem of sustainable development of the climate–nature–society system (CNSS) taking into account both natural and demographic processes. The complexity of the Big data from CNSS is that satellite environmental monitoring does not provide data that can help to assess the CNSS characteristics with high reliability. The second complexity of the big data that was observed by the researchers is the non-inclusion of biogeochemical, biocenotic, hydrophysical, climatic, and socioeconomic processes. Each processes have their own variability because of the mode of data generation. The Earth Observing System Data and Information System (EOSDIS) that was used by the researcher have volume growth of approximately 8.5 TB daily. The existing tool for treating big data in CNSS is the CNSS global simulation model (CNSSGSM) that works on two mathematical principles, that is, balance equations of fluxes of matters and evolutionary algorithm that resolve the balance model. Figure 1.10 shows those complexities in the CNSS-GSM. The authors were successful in improving the CNSS operation using simple biosphere models and complex simulation models that requires big data processing.

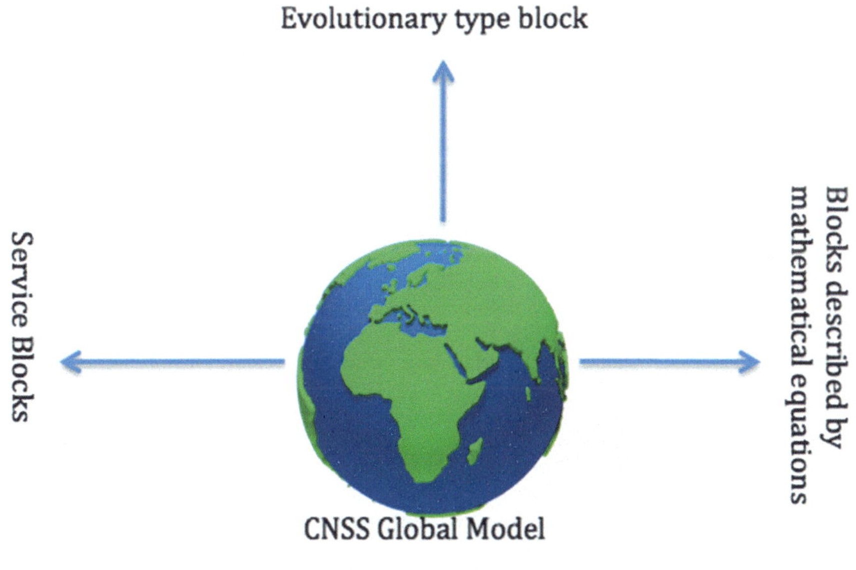

Fig. 1.10 The structure of the CNSSGSM [44]

References

1. Emetere ME (2016) Numerical modeling of West Africa regional scale aerosol dispersion. A doctoral thesis submitted to Covenant University, Nigeria, pp 1–289
2. WHO (2018a) Air pollution. http://www.who.int/airpollution/en/. Accessed on 5th July 2018
3. Ritchie H, Roser M (2018) Air pollution. Published online at Our World in Data.org. Retrieved from: https://ourworldindata.org/air-pollution. Accessed on 5th July 2018
4. WHO (2018b) Drinking water. http://www.who.int/news-room/fact-sheets/detail/drinking-water. Accessed on 5th July 2018
5. Guha-Sapir D, Hoyois P, Wallemacq P, Below R (2018) Annual disaster statistical review 2016. https://emdat.be/sites/default/files/adsr_2016.pdf. Accessed on 23 June 2018
6. EMDAT (2017) OFDA/CRED international disaster database. Universite de Louvain, Brussels Belgium
7. Emetere ME (2014) Volcanic eruption trends in the five-years pre-eruption era. J Volcanol Seismolog 8(6):411–417
8. Emetere ME (2017) Monitoring the 3-year thermal signatures of the Calbuco pre-volcano eruption event. Arab J Geosci 10:94. https://doi.org/10.1007/s12517-017-2861-z
9. Emetere ME (2016) Statistical examination of the aerosols loading over Mubi-Nigeria: the satellite observation analysis. Geographica Panonica 20(1):42–50
10. Emetere ME (2016) Generation of atmospheric constants over some locations in West Africa: a theoretical aid for measuring instruments design. Int J Eng Res Afr 27:119–146
11. DBW (2016) Environmental modeling. https://www.designingbuildings.co.uk/wiki/Environmental_modeling. Accessed on 7th July 2018
12. Jolankai G (1997) Description of the CAL programme on water quality modeling version 1.1 basic river water quality models. http://unesdoc.unesco.org/images/0012/001213/121363Eo.pdf
13. Udeze C (2018) 10 horrible effects of oil spill in Nigeria. https://buzznigeria.com/oil-spill/. Accessed on 10th July 2018
14. Ojewumi ME, Emetere ME, Babatunde DE, Okeniyi JO (2017) In situ bioremediation of crude petroleum oil polluted soil using mathematical experimentation. Int J Chem Eng 11
15. Sanni SE, Emetere M (2016) Mathematical modelling of insitu-bioremediation of crude oil polluted soil, Sci Eng Appl 1(4):27–32
16. McGill (2018) Environmental engineering. https://www.mcgill.ca/civil/undergrad/areas/environmental. Accessed on 7th July 2018
17. Tam VW-Y, and Lu W (2016) Construction Waste management profiles, practices, and performance: a cross-jurisdictional analysis in four countries, sustainability. Sustainability 8:190
18. Lu W, Yuan HP (2011) A framework of understanding waste management studies in construction. Waste Manag 31:1252–1260
19. Enríquez-de-Salamanca A, Martín-Aranda RM, Díaz-Sierra R (2016) Consideration of climate change on environmental impact assessment in Spain. Environ Impact Assess Rev 57:31–39
20. Geneletti D, Beinat E, Chung CF, Fabbri AG, Scholten HJ (2000) Accounting for uncertainty factors in biodiversity impact assessment: lessons from a case study. Environ Impact Assess Rev 23(4):471–487
21. Schaubroeck T, Deckmyn G, Giot O, Campioli M, Vanpoucke C, Verheyen K, Rugani B, Achten W, Verbeeck H, Dewulf J, Muys B (2016) Environmental impact assessment and monetary ecosystem service valuation of an ecosystem under different future environmental change and management scenarios; a case study of a scots pine forest. J Environ Manage 173:79–94
22. Nature (2018) Environmental chemistry. https://www.nature.com/subjects/environmental-chemistry. Accessed on 7th July 2018
23. Øystein Hov, Terblanche D, Carmichael G, Jones S, Ruti PM, Tarasova O (2017) Five priorities for weather and climate research. https://www.nature.com/articles/d41586-017-08463-3. Accessed on 7th July 2018

24. NCAR (2018) Volcanic emissions: more than meets the eye? new method estimates climatic influence of eruptions. https://www2.ucar.edu/atmosnews/just-published/20003/volcanic-emissions-more-meets-eye. 10th July 2018
25. GIG (2018) Optimal insulation for desert areas. http://www.gig-group.com/en/newspaper/article/3004. 10th July 2018
26. NASA (2018) NASA maps shed light on carbon dioxide's global nature. https://phys.org/news/2008-10-nasa-carbon-dioxide-global-nature.html. Accessed on 10th July 2018
27. Yukimoto S, Noda A, Kitoh A, Hosaka M, Yoshimura H, Uchiyama T, Shibata K, Arakawa O, Kusunoki S (2006) Present-day climate and climate sensitivity in the meteorological Re-search Institute coupled GCM version 2.3 (MRI-CGCM2.3). J Meteor Soc Japan 84:333–363
28. Stier P, Feichter J, Kinne S, Kloster S, Vignati E, Wilson J, Ganzeveld L, Tegen I, Werner M, Balkanski Y, Schulz M, Boucher O, Minikin A, Petzold A (2005) The aerosol-climate model ECHAM5-HAM. Atmos Chem Phys 5:1125–1156
29. Matsui H (2017) Development of a global aerosol model using a two-dimensional sectional method: 1. Model design. J Adv Model Earth Syst 9:1921–1947
30. Carneva C, Decanini E, Volta M (2008) Design and validation of a multiphase 3D model to simulate tropospheric pollution. Sci Total Environ 390(1):166–176
31. Turco RP, Hamill P, Toon OB, Whitten RC, Kiang CS (1979) A one-dimensional model describing aerosol formation and evolution in the stratosphere: I. Physical processes and mathematical analogs. J Atmos Sci 36:699–717
32. Toon OB, McKay CP, Ackerman TP, Santhanam K (1989) Rapid calculation of radiative heating rates and photo dissociation rates in inhomogeneous multiple scattering atmospheres. J Geophys Res 94(D13):16287–16301
33. Spracklen DV, Pringle KJ, Carslaw KS, Chipperfield MP, Mann GW (2005) A global off-line model of size-resolved aerosol microphysics: I. Model development and prediction of aerosol properties. Atmos Chem Phys 5:2227–2252
34. RAMS (2018) http://rams.atmos.colostate.edu/rams-description.html. Accessed on 7th July 2018
35. Saleeby SM, van den Heever SC (2013) Developments in the CSU-RAMS aerosol model: emissions, nucleation, regeneration, deposition, and radiation. J Appl Meteor Climatol 52:2601–2622
36. Igel AL, van den Heever SC, Johnson JS (2018) Meteorological and land surface properties impacting sea breeze extent and aerosol distribution in a dry environment. J Geophys Res: Atmos 123(1):22–37
37. Bessagnet B, Rosset B (2001) Fractal modeling of carbonaceous aerosols—application to car exhaust plumes. Atmos Environ 35:4751–4762
38. Van Velthoven PFJ, Kelder H (1996) Estimates of stratosphere-troposphere exchange: sensitivity to model formulation and horizon-tal resolution. J Geophys Res 101:1429–1434
39. Guillaume B, Liousse C, Rosset R, Cachier H, Van Velthoven P, Bessagnet B, Poisson N (2007) ORISAM-TM4: a new global sectional multi-component aerosol model including SOA formation—focus on carbonaceous BC and OC aerosols. Tellus B: Chem Phys Meteorol 59(2):283–302
40. Debry E, Fahey K, Sartelet K, Sportisse B, Tombette M (2007) Technical note: a new SIze REsolved aerosol model (SIREAM). Atmos Chem Phys 7:1537–1547
41. Sas (2018) Big data: what it is and why it matters. https://www.sas.com/en_us/insights/big-data/what-is-big-data.html. Accessed on 19th July 2018
42. Ponce Romero JM, Hallett SH, Jude S (2017) Leveraging big data tools and technologies: addressing the challenges of the water quality sector. Sustainability 9:2160
43. Jay H (2017). Environmental analysis in the era of cloud and big data platforms. https://www.ibm.com/blogs/bluemix/2017/01/environmental-analysis-era-cloud-big-data-platforms/. Accessed on 20th July 2018
44. Varotsos AC, Krapivin VF (2017) A new big data approach based on geoecological information-modeling system. Big Earth Data 1(1–2):47–63

Chapter 2
Numerical Methods

2.1 Introduction

The role of mathematics and physics in understanding the basic principles of environmental models cannot be over-emphasized. The role of climate change and new dimension of theoretical inclusions in fields of environmental study are salient reason why an easier and approximate approach is relevant. The general believe before now mandates environmental researchers to possess a degree of accuracy in the use of analytical technique. In recent times, the applications of mathematical software have made it easier for very cumbersome equations to be solved. Hence, it is easier and faster to solve more environmental problems numerically. In this chapter, background, focus and application of numerical methods were highlighted with worked examples.

2.2 Background to Numerical Methodology

Numerical method is a mathematical tool (existing or newly developed) designed to solve numerical problems. Numerical problems may be series of events connected to each other. What are the types of model that requires the use of numerical methods? There are many models but the authors have mentioned a few for illustration purposes. The models are:

i. Physical Models—are framework of ideas and concepts from which the user interprets various observations and experimental results. It is a simplified material representation, usually on a reduced scale, of an object or phenomenon that needs to be investigated.
ii. Models of natural science—are framework that examine 'exact science' or a phenomenon based hypothesis or known scientific facts.

M. E. Emetere and E. T. Akinlabi, *Introduction to Environmental Data Analysis and Modeling*, Lecture Notes in Networks and Systems 58,
https://doi.org/10.1007/978-3-030-36207-2_2

iii. Mathematical models—are framework that translates problems noticed in systems into tractable mathematical formulations whose theoretical and numerical analysis show high accuracy of the system operation. The process of developing a mathematical model is termed mathematical modeling.
iv. Engineering Models—are framework that makes use of physical and mathematical model on a higher abstraction level to provide reliable and accurate solutions.
v. Environmental models—are framework that make use of multidisciplinary knowledge (mathematical, computational, quantitative, animation etc.) to explain, explore and predict the Earth's response to environmental change, both natural and human-induced.

Recall, that the focus of this book is on the environmental model. The pictorial chat describes the role of numerical methods in environmental modeling (Fig. 2.1).

In numerical methods, three parameters are very salient i.e. mathematical definition and formulation, consistency and convergence. The processes of using numerical methods to solve problems are referred to as numerical analysis. The implementation of a numerical method with an appropriate programming language is referred to as numerical algorithm. Within numerical methods, there is subdivision of methods used to solve problems. Below are some of the methods and what they represent

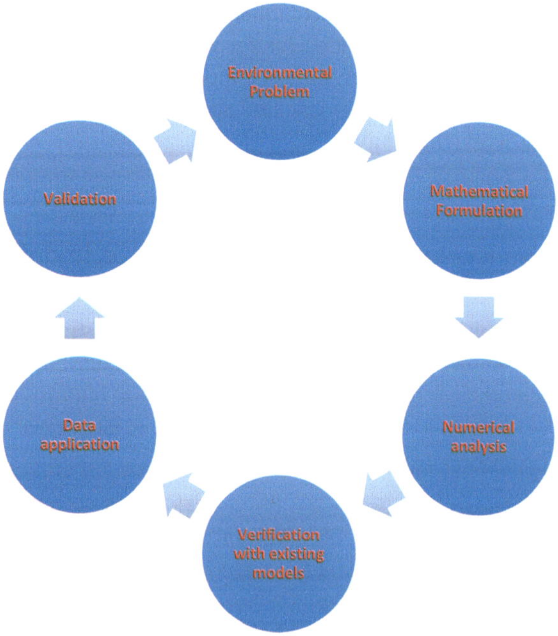

Fig. 2.1 Role of numerical methods in environmental modeling

i. Bisection method is a root-finding method that repeatedly bisects an interval and then selects a subinterval in which a root must lie for further processing. The bisection method is the easiest to numerically implement considering environmental data that has high susceptibility to variation e.g. dataset on solar irradiation, precipitation, river contamination aerosol optical depth, aerosol loading etc. The main disadvantage is that convergence is slow. If the bisection method results in a computer program that runs too slow, then other faster methods maybe chosen; otherwise it is a good choice of method.
ii. Newton's method (also known as the Newton–Raphson method), is a method for finding successively better approximations to the roots (or zeroes) of a real-valued function. This is one of the very fast method. However, the challenge of this method in practical terms is that it requires analytical computation of the derivative and the method may not always converge to the desired root. For environmental modeller or scientist, he is not concerned with the convergence because the computer program is always aimed at fitting the dataset trends.
iii. Secant method is a root-finding algorithm that uses a succession of roots of secant lines to better approximate a root of a function f. The secant method can be thought of as a finite-difference approximation of Newton's method. The Secant Method is also a very fast method like the Newton's Method, and is used when a faster convergence than Bisection is desired. Its shortcoming is that it is too difficult or impossible to take an analytical derivative of the function f.
iv. Gaussian Elimination becomes relevant when the mathematical transformation of the problem results into systems of equation. Gaussian elimination (also known as row reduction) is a sequence of operations performed on the corresponding matrix of coefficients. This method is regarded as the standard numerical algorithm to solve a system of linear equations. Bates et al. [1] used the Gaussian elimination method in developing a model that will gain computational efficiency solve solar fluxes at each waveband by solving a penta-diagonal matrix. The calculation was able to improve on the uncertainty in DCF by a factor of 25% less than the "structural uncertainties" used in the IPCC-2001 global estimates of direct aerosol climate forcing. Like the Gaussian elimination method, the Gauss-Jordan elimination is one way to solve linear systems. Gauss-Jordan elimination involves creating an augmented matrix of both sides of the developed or derived equations, changing this matrix into reduced row echelon form, and then finishing up the problem to find the required solution.
v. Interpolation method is a method of constructing new data points within the range of a discrete set of known data points. In environmental studies, the number of data points obtained within dataset or experimentation is used to represent the values of a function for a limited number of values of the independent variable. The different cases of interpolation include: interpolation with unevenly spaced points e.g. Lagrange interpolation and Newton's divided difference interpolation; interpolation with Evenly Spaced Points e.g. Newton's forward difference interpolation formula, and Newton's backward difference interpolation formula; and Spline Interpolation and Cubic Splines.

vi. Iterative method is a mathematical procedure that uses an initial guess to generate a sequence of improving approximate solutions for a class of problems, in which the n-th approximation is derived from the previous ones. In writing a computational program to solve either differential or integral problems that originate from an environmental problem, the iterative method is the best method because it helps the modeller to see the convergence or consistency of the solution.

vii. Power Method is very good at approximating the external eigenvalues of the matrix, that is, the eigenvalues having largest and smallest module, denoted by λ_1 and λ_n respectively, as well as their associated eigenvectors. The algorithm for solving the power method is also known as the Von Mises iteration.

viii. Numerical differentiation is the process of finding the numerical value of a derivative of a given function at a given point. It describes algorithms for estimating the derivative of a mathematical function or function subroutine using values of the function and perhaps other knowledge about the function. Most model formulation is based on numerical differentiation. The most routine for solving numerical differentiation is the methods based on finite differences. The branches of finite differential method are derivatives using Newton's forward difference formula, derivatives using Newton's backward difference formula, and derivatives using Newton's divided difference formula. Also, another routine of numerical differential is the initial value problems. Methods for solving the initial value problems are Taylor series method, modified Euler and Heun's methods, Runge-Kutta methods, Taylor series method, Runge-Kutta fourth order method, multi-step methods, Predictor-Corrector methods, Corrector methods, Adams-Moulton methods, Milne-Simpson methods, Predictor-Corrector methods.

ix. Numerical quadrature refers to any method for numerical approximation of the value of a definite integral. The goal is to attain a given level of precision with the fewest possible function evaluations. The subdivision of numerical quadrature is namely, polynomial quadrature formulas, Gauss quadrature, composite quadrature, and adaptive quadrature.

x. Numerical integration is the approximate computation of an integral using numerical techniques. It constitutes a broad family of algorithms for calculating the numerical value of a definite integral, and by extension, the term is also sometimes used to describe the numerical solution of differential equations. The subdivision method for solving numerical integration using Trapezium rule, Simpson's 1/3 rule, Simpson's 3/8 rule, Romberg method, Gauss-Legendre integration rules, evaluation of double integrals using trapezium rule, evaluation of double integrals by Simpson's rule.

2.3 General Outline on Numerical Methods

This section is design to take readers through a summarized documentary of some numerical methods that will be used to process dataset and analyse big data. The outline for discussion is tailored towards succeeding chapters.

2.3.1 *Root Finding*

Root finding technique is essential when an explicit analytical solution is impossible e.g. f(x) = 0. Determining the roots/zeros of an equation is a problem of great importance in science and engineering. The two types of equation in root finding is the algebraic and transcendental equation. The difference between both equation is that in algebraic equation an algebraic term is equal to zero e.g. $7x^4+3x^3+2x^2+x-5=0$, while the transcendental equation is an equation which comprise of exponential function, logarithmic function and trigonometry functions e.g. $ye^{4y}+\sin^2(y)=0$. The solution of any of the equation can be a simple root or multiple roots. A number (y) is regarded as a simple root if when substituted as x in f(x) = 0 gives f(y) = 0 and f′(y) ≠ 0. A number (z) is regarded as a multiple root if when substituted as x in f(x) = 0 gives f(z) = 0, f′(z) = 0, and f″(z) ≠ 0.

Example 1—Simple root
Find the root of $f(x)=7x^4+3x^3+2x^2+x-5=0$.

Solution
Let x = −1, then,

$$f(-1)=7(1)+3(-1)+2(1)+(-1)-5=0$$

however,

$$f'(x)=28x^3+9x^2+4x+1=0$$
$$f'(x)=28(-1)+9(1)+4(-1)+1=-22$$

Hence,

$$f'(-1)\neq 0.$$

The simple root is −1.

Example 2—Multiple root
Find the root of $f(x)=3x^3-2x^2-28x+40=0$.

Solution
Let x = 2, then,

$$f(2) = 3(8) - 2(4) - 28(2) + 40 = 0$$

however,

$$f'(x) = 9x^2 - 4x - 28 = 0$$
$$f'(2) = 9(4) - 4(2) - 28 = 0$$

Also,

$$f''(x) = 18x - 4 = 0$$
$$f'(2) = 18(2) - 4 = 32$$

Hence,

$$f''(2) \neq 0.$$

The multiple root is 2.

The above examples are very trivial because they are regarded as direct method. Most time, mathematical formulation of an event may be complex and requires a complex method of finding its roots e.g. iterative method. The iterative method works on the idea of successive approximations. This begins by slotting one or two initial approximations to the root into a specified formula to obtain sequence of approximations as its outcome. In this section, three methods for solving root iteratively were considered i.e. Bisection method, Newton's method and Secant method. The definitions of these methods have been provided in the last section.

The bisection method is known with the formula below:

$$z_2 = z_0 + \frac{z_1 - z_0}{2}$$

where z_0 are z_1 are the initial approximations. So if you consider two extremes of z_0 and z_1, the increment between the two extreme is the average i.e. $\frac{z_1 - z_0}{2}$. In the next iteration, the bisection representation becomes:

$$z_3 = z_1 + \frac{z_2 - z_1}{2}$$
$$z_4 = z_2 + \frac{z_3 - z_2}{2}$$

The Newton's method is defined by the formula below:

$$z_{n+1} = z_n - \frac{f(z_n)}{f^{\iota}(z_n)}$$

where z_n is the initial approximation in all the iterative processes.

The Secant method is defined by the formula below:

$$z_{n+1} = z_n - \frac{(z_n - z_{n-1})f(z_n)}{f(z_n) - f(z_{n-1})}$$

where z_{n-1} and z_n are the initial approximations.

Example 3—Bisection method
Find $\sqrt{2.743}$ using the Bisection method.

Solution
The success of solving any of the iterative method is taking a good guess from the beginning. It saves time and reduces bogus outcomes. In this case, the guess is $z_0 = 1$, $z_1 = 2$.

$$z_2 = z_0 + \frac{z_1 - z_0}{2}$$

$$z_2 = 1 + \frac{2-1}{2} = 1.5$$

$$z_3 = z_1 + \frac{z_2 - z_1}{2}$$

$$z_3 = 2 + \frac{1.5-2}{2} = 1.75$$

$$z_4 = z_2 + \frac{z_3 - z_2}{2}$$

$$z_4 = 1.5 + \frac{1.75-1.5}{2} = 1.625$$

$$z_5 = z_3 + \frac{z_4 - z_3}{2}$$

$$z_5 = 1.75 + \frac{1.625-1.75}{2} = 1.6875$$

$$z_6 = z_4 + \frac{z_5 - z_4}{2}$$

$$z_6 = 1.625 + \frac{1.6875-1.625}{2} = 1.65625$$

$$z_7 = z_5 + \frac{z_6 - z_5}{2}$$

$$z_7 = 1.6875 + \frac{1.65625-1.6875}{2} = 1.65625$$

$$z_8 = z_6 + \frac{z_7 - z_6}{2}$$

$$z_8 = 1.65625 + \frac{1.65625 - 1.65625}{2} = 1.65625$$

Note that the sequence of iterates converged to the exact root 1.65625.

As shown in Example 3, one the signs that you have reached the root of a function is when it converges to a particular number.

Example 4—Newton's method

Find $\sqrt{3}$ using the Newton's method when $f(x) = x^2 - 3$.

Solution

Recall that the formula for Newton's method is given as:

$$z_{n+1} = z_n - \frac{f(z_n)}{f^{\prime}(z_n)}$$

let $z_n = 1,\ f^{\prime}(z_n) = 2x$

$$z_1 = 1 - \frac{-2}{2} = 2$$
$$z_2 = 2 - \frac{1}{4} = 1.75$$
$$z_3 = 1.75 - \frac{0.0625}{3.5} = 1.73214$$

Example 5—Newton's method

Do the iterations converge for finding 1/2R, where N = 15. Hence, find 1/13, using 0.025 as initial approximation?

From the above, let $z_n = \frac{1}{2R}$, $2R = \frac{1}{z_n}$ so that $f(z_n) = 2R - \frac{1}{z_n}$, $z_n = 0.25$

$$f^{\prime}(z_n) = \frac{1}{z_n^2}$$

$$z_{n+1} = z_n - \frac{f(z_n)}{f^{\prime}(z_n)} = z_n - \frac{2R - \frac{1}{z_n}}{\frac{1}{z_n^2}}$$

$$z_{n+1} = z_n - 2Rz_n^2 + z_n = 2z_n - 2Rz_n^2$$
$$z_1 = 2(0.025) - 2 * 15(0.025)^2 = 0.03125$$
$$z_2 = 2(0.03125) + 2 * 15(0.03125)^2 = 0.0332$$
$$z_3 = 2(0.0332) + 2 * 15(0.0332)^2 = 0.0333$$
$$z_4 = 2(0.0333) + 2 * 15(0.0333)^2 = 0.0333$$

Note that the sequence of iterates converged to the exact root 0.0333.

The second aspect of this section is to introduce computational angle to resolving root finding problems. The computer language that is used is the C++.

Example 6—Newton's method (Codes)
Find $\sqrt{3}$ using the Newton's method when $f(x) = x^2 - 3$.

Solution
The c++ codes below (in purple) would run the same command to give the convergence of the iteration.

```
void bisec()
{
Int_t n = 30;
double x1=1;
double s;
double fn;
double fo;


do
{
        fn=x1*x1-3;
        fo=2*x1;
s = x1 - (fn / fo) ;
x1=s;
        out_data<<s<<" "<< fn<<" "<< fo<<" "<<endl;
}
while (x1 == s);


out_data<<s<<" "<< x1<<endl;
return 0;


}
```

Solution

No.	z_n	$f(z_n)$	$f'(z_n)$
1	2.000	−2.000	2.000
2	1.75	1.000	4.000
3	1.73214	0.0625	3.5
4	1.73205	0.000318878	3.46429
5	1.73205	8.47267e−09	3.4641

At the fifth term the convergence was confirmed.

Example 7—Newton's method (Codes)
Using the Newton's method find the root of $f(x) = x^5 + 5x^3 + 7x^2 - 3$.

Solution

```
void bisec()
{
Int_t n = 30;
double x1=1;
double s;
double fn;
double fo;

do
{
      fn=x1*x1*x1*x1*x1+5*x1*x1*x1+7*x1*x1-3;
      fo=5*x1*x1*x1*x1+15*x1*x1+14*x1;
s = x1 - (fn / fo) ;
x1=s;
      out_data<<s<<" "<< fn<<" "<< fo<<" "<<endl;
 }
 while (x1 == s);

 out_data<<s<<" "<< x1<<endl;
 return 0;

 }
```

No.	z_n	$f(z_n)$	$f'(z_n)$
1	0.705882	10.000	34.000
2	0.575666	2.42174	18.5978
3	0.550863	0.336808	13.5793
4	0.550024	0.0106663	12.7242
5	0.550023	1.18951e−05	12.6959
6	0.550023	1.48486e−11	12.6958
7	0.550023	4.44089e−16	12.6958
8	0.550023	4.44089e−16	12.6958

Example 8—Newton's method (Codes)
Find the smallest negative root in magnitude of the equation $f(z_n) = 3^x + 3x^2$.

Solution
The differentiation of the $f(z_n)$ is given as $f'(z_n) = 3^x \log(3) + 6x$

```
void bisec()
{
Int_t n = 30;
double x1=1;
double s;
double fn;
double fo;

do
{
        fn=pow(3,x1)+3*pow(x1,2);
        fo=pow(3,x1)*log(3)+6*x1;
 s = x1 - (fn / fo) ;
 x1=s;
        out_data<<x1<<" "<< fn<<" "<< fo<<" "<<endl;
 }
 while (x1 == s);

 out_data<<x1<<" "<< fn<<" "<< fo<<" "<<endl;
 return 0;

 }
```

Solution 8a (when $z_n = 1$)

No.	z_n	$f(z_n)$	$f'(z_n)$
1	0.35455	6.00	9.29584
2	−0.139798	1.85338	3.74914
3	−8.99997	0.916259	0.103413
4	−4.49998	242.998	−53.9997

Solution 8b (when $z_n = 0.5$)

No.	z_n	$f(z_n)$	$f'(z_n)$
1	−0.00624629	2.48205	4.90285
2	−0.948974	0.993278	1.05362
3	−0.373417	3.05421	−5.30653
4	−0.373417	3.05421	−5.30653

It can be observed that the negative root in solution (8a) and (8b) differs. Also, solution (8b) converged to negative root −0.373417 while solution (8a) did not converged. This gives more credence to the earlier assertion that the choice of the initial approximation determines the success and accuracies of the root finding method.

2.3.2 System of Equations

In this section, the trivial cases are considered before the complex cases. First the Gaussian elimination method was considered. The principle of this method is to reduce the lower rows of the matrices such that the solution of the first row would depend on the lowest row of the matrices.

Example 9—Gaussian Elimination method
Solve the equations below using the Gaussian elimination method.

$$
\begin{aligned}
&5z_1 - 3z_2 + 7z_3 - 2z_4 = -8 \\
&-10z_1 + 3z_2 + 9z_3 - 7z_4 = -12 \\
&10z_1 - 4z_2 + 5z_3 - 5z_4 = -18 \\
&-5z_1 + 5z_2 - 3z_3 - 4z_4 = -10
\end{aligned}
$$

Solution
The matrix of the above equations can be written as:

$$\begin{pmatrix} 5 & -3 & 7 & -2 & -8 \\ -10 & 3 & 9 & -7 & -12 \\ 10 & -4 & 5 & -5 & -18 \\ -5 & 5 & -3 & -4 & -10 \end{pmatrix}$$

the processes for elimination is indicated in red. The first process is to eliminate the first column of the second, third and fourth rows.

$$\left.\begin{pmatrix} 5 & -3 & 7 & -2 & -8 \\ -10 & 3 & 9 & -7 & -12 \\ 10 & -6 & 5 & -5 & -18 \\ -5 & 5 & -3 & -4 & -10 \end{pmatrix}\right| \begin{matrix} \boldsymbol{R}_1 \\ \boldsymbol{R}_2 \\ \boldsymbol{R}_3 \\ \boldsymbol{R}_4 \end{matrix}$$

$$\left.\begin{pmatrix} 5 & -3 & 7 & -2 & -8 \\ 0 & -3 & 23 & -11 & -28 \\ 0 & 0 & -9 & 1 & -2 \\ 0 & 2 & 4 & -6 & -18 \end{pmatrix}\right| \begin{matrix} \boldsymbol{R}_1 \\ \boldsymbol{2R}_1 + \boldsymbol{R}_2 \\ \boldsymbol{R}_3 - \boldsymbol{2R}_1 \\ \boldsymbol{R}_1 + \boldsymbol{R}_4 \end{matrix}$$

The second process is to eliminate the second column of the third and fourth rows.

$$\left.\begin{pmatrix} 5 & -3 & 7 & -2 & -8 \\ 0 & -3 & 23 & -11 & -28 \\ 0 & 0 & -9 & -1 & -2 \\ 0 & 0 & 58 & -40 & -110 \end{pmatrix}\right| \begin{matrix} \boldsymbol{R}_1 \\ \boldsymbol{R}_2 \\ \boldsymbol{R}_3 \\ \boldsymbol{2R}_2 + \boldsymbol{3R}_4 \end{matrix}$$

The third process is to eliminate third column of the fourth row.

$$\left.\begin{pmatrix} 5 & -3 & 7 & -2 & -8 \\ 0 & -3 & 23 & -11 & -28 \\ 0 & 0 & -9 & -1 & -2 \\ 0 & 0 & 0 & -418 & -1106 \end{pmatrix}\right| \begin{matrix} \boldsymbol{R}_1 \\ \boldsymbol{R}_2 \\ \boldsymbol{R}_3 \\ \boldsymbol{58R}_3 + \boldsymbol{9R}_4 \end{matrix}$$

The final elimination matrix is:

$$\begin{pmatrix} 5 & -3 & 7 & -2 & -8 \\ 0 & -3 & 23 & -11 & -28 \\ 0 & 0 & -9 & -1 & -2 \\ 0 & 0 & 0 & -418 & -1106 \end{pmatrix}$$

$$5z_1 - 3z_2 + 7z_3 - 2z_4 = -8$$
$$-3z_2 + 23z_3 - 11z_4 = -28$$

$$-9z_3 - z_4 = -2$$
$$-418z_4 = -1106$$

From the above equation, the solution of each term can be written as:

$$z_4 = 2.6459, z_3 = 0.07177, z_2 = 29.3554, \text{ and } z_1 = 16.97113.$$

Example 9—Gaussian Elimination method (Codes)
Solve the equations below using the Gaussian elimination method.

$$4z_1 - 4z_2 + 7z_3 - 2z_4 = 8$$
$$10z_1 + 3z_2 + 5z_3 - 7z_4 = 3$$
$$3z_1 - 4z_2 + 2z_3 - z_4 = 5$$
$$4z_1 + 5z_2 - z_3 - 3z_4 = 2$$

The c++ code used in this section is adopted from Thoma [2].

```
#include <iostream>

#include <cmath>

#include <vector>


using namespace std;


void print(vector< vector<double>> A) {
unsigned long n = A.size();
  for (int i=0; i<n; i++) {
    for (int j=0; j<n+1; j++) {
      cout << A[i][j] << "\t";
      if (j == n-1) {
        cout << "| ";
      }
    }
    cout << "\n";
  }
  cout << endl;
```

```
}

vector<double> gauss(vector< vector<double>> A) {
unsigned long n = A.size();

   for (int i=0; i<n; i++) {
      // Search for maximum in this column
      double maxEl = abs(A[i][i]);
      int maxRow = i;
      for (int k=i+1; k<n; k++) {
         if (abs(A[k][i]) > maxEl) {
            maxEl = abs(A[k][i]);
            maxRow = k;
         }
      }

      // Swap maximum row with current row (column by column)
for (int k=i; k<n+1;k++) {
         double tmp = A[maxRow][k];
         A[maxRow][k] = A[i][k];
         A[i][k] = tmp;
      }

      // Make all rows below this one 0 in current column

      for (int k=i+1; k<n; k++) {
         double c = -A[k][i]/A[i][i];
         for (int j=i; j<n+1; j++) {
            if (i==j) {
               A[k][j] = 0;
            } else {
               A[k][j] += c * A[i][j];
            }
         }
      }
   }
```

```
    // Solve equation Ax=b for an upper triangular matrix A
    vector<double> x(n);
    for (int i= 0; i<n; i++) {
       x[i] = A[i][n]/A[i][i];
       for (int k= 0; k<n; k++) {
          A[k][n] -= A[k][i] * x[i];
       }
    }
    return x;
 }
   int main() {
  int n;
  cin >> n;

  vector<double> line(n+1,0);
vector< vector<double>> A(n,line);

  // Read input data
  for (int i=0; i<n; i++) {
     for (int j=0; j<n; j++) {
        cin >> A[i][j];
     }
  }

  for (int i=0; i<n; i++) {
     cin >> A[i][n];
  }

  // Print input
  print(A);
```

```
    // Calculate solution
    vector<double> x(n);
    x = gauss(A);
    // Print result
    cout << "Result:\t";
    for (int i= 0; i<n; i++) {
       cout << x[i] << " ";
    }
    cout << endl;
}
```

Solution

$$\begin{array}{cccc|c} 4 & -4 & 7 & -2 & 5 \\ 8 & 10 & 3 & 5 & -1 \\ -7 & 3 & 3 & -4 & -3 \\ 2 & -1 & 5 & 4 & 2 \end{array}$$

$z_4 = -0.1225$, $z_3 = -0.329787$, $z_2 = 0.258134$, and $z_1 = -0.0802407$.

2.3.3 Numerical Differentiation

In this section, the numerical differentiation that would be used in succeeding chapters of this book was used.

Example 10—Numerical Differentiation method (Codes)
Solve the system of equations using any convenient boundary or initial conditions

$$\frac{\partial x}{\partial t} = \frac{1}{\rho}\frac{\partial z}{\partial t} + \frac{1}{\varphi}\frac{\partial y}{\partial t} + \frac{V_x * c_t}{p} + \frac{V_z * c_t}{s}$$
$$\frac{\partial y}{\partial t} = \varphi\frac{\partial x}{\partial t} - \beta\frac{\partial z}{\partial t} + V_y + \frac{V_y * c_t}{p} + \frac{V_y * c_t}{s}$$
$$\frac{\partial z}{\partial t} = \rho\frac{\partial x}{\partial t} - \frac{1}{\beta}\frac{\partial y}{\partial t} + V_z + \frac{V_z * c_t}{p} + \frac{V_z * c_t}{s}$$

The parameters are $V_y = 18$; $V_x = 18$; $V_z = 9$; p = 100; s = 40; $c_t = 15$; $\rho = \frac{V_y}{V_x}$; $\beta = \frac{V_y}{V_z}$; $\rho = \frac{V_z}{V_x}$.

Solution
The c++ code used to solve the system of differential equation. For beginner, it

is advisable to download and compile the 'Boost' library before running the code provided below.

```
#include <iostream>
#include <boost/array.hpp>

#include <boost/numeric/odeint.hpp>


using namespace std;
using namespace boost::numeric::odeint;
const double vy = 18;
const double vx = 18;
const double vz = 9;
const double p = 100;
const double s = 40;
const double ct = 15;
const double sigma = vy/vx;
const double beta = vy/vz;
const double row = vz/vx;
typedef boost::array< double , 3 > state_type;

void lorenz( const state_type &x , state_type &dxdt , double t )
{
dxdt[0] =  x[2]/row + x[1]/sigma +(vx*ct)/p +(ct*vz)/s;
dxdt[1] = sigma * x[0] - beta * x[2] + vy + (ct*vy)/p- (ct*vy)/s ;
dxdt[2] = row * x[0] -x[1]/beta +vz+(ct*vz)/p-(ct*vz)/s;
}
void write_lorenz( const state_type &x , const double t )
{
cout<< t << '\t' << 1/x[0] << '\t' << 1/x[1] << '\t' << 1/x[2] << endl;
}
int main(int argc, char **argv)
{
state_type x = { 1 , 1 , 1 }; // initial conditions
   integrate( lorenz , x , 0.0 , 10.0 , 0.1 , write_lorenz );
}
```

t	$\frac{\partial x}{\partial t}$	$\frac{\partial y}{\partial t}$	$\frac{\partial z}{\partial t}$
0.000	1.000	1.000	1.000
0.1	2.0408	2.27559	1.69038
0.217032	3.59368	3.74482	2.4936
0.347232	5.74672	5.39684	3.40599
0.49076	8.65637	7.29772	4.46564
0.648082	12.5352	9.54876	5.73031
0.819847	17.6791	12.296	7.28309
1.00657	24.498	15.7419	9.23907
1.20824	33.5473	20.1585	11.753
1.4239	45.5532	25.8979	15.0256
1.65147	61.428	33.3991	19.3068
1.88807	82.29	43.1958	24.9012
2.12468	108.747	55.5813	31.9757
2.37017	143.705	71.9203	41.3098
2.61567	188.401	92.7948	53.2356
2.86116	245.54	119.471	68.4764
3.10665	318.583	153.566	87.9559
3.35215	411.952	197.145	112.854
3.59764	531.303	252.849	144.68
3.84313	683.865	324.051	185.361
4.08863	878.878	415.065	237.362
4.33412	1128.15	531.404	303.831
4.57962	1446.79	680.113	388.795
4.82511	1854.09	870.202	497.401
5.0706	2374.72	1113.18	636.227
5.3161	3040.22	1423.77	813.681
5.56159	3890.9	1820.79	1040.51
5.80708	4978.28	2328.27	1330.46

(continued)

(continued)

t	$\frac{\partial x}{\partial t}$	$\frac{\partial y}{\partial t}$	$\frac{\partial z}{\partial t}$
6.05258	6368.22	2976.96	1701.09
6.29807	81,44.93	3806.16	2174.84
6.54357	10,416,000	4866.07	2780.42
6.78906	13,319,000	6220.92	3554.5
7.03455	17,029.8	7952.75	4543.97
7.28005	21,773.1	10,166.5	5808.76
7.52554	27,836.2	12,996.2	7425.48
7.77104	35,586.4	16,613.2	9492.07
8.01653	45,493.2	21,236.7	12133.7
8.26202	58,156.5	27,146.7	15,510.3
8.50752	74,343.4	34,701.2	19,826.5
8.75301	95,034.3	44,357.8	25,343.7
8.9985	121,483,000	56,701.3	32,396.1
9.244	155,290,000	72,479.4	41,410.8
9.48949	198,505,000	92,647.8	52,933.9
9.73499	253,744,000	118,428,000	67,663.3
9.98048	324,353,000	151,382,000	86,491.2
10,000	330,748,000	154,366,000	88,196.2

Example 10—Numerical Differentiation method (Codes)
Solve the system of equations using any convenient boundary or initial conditions

$$\left[\frac{\partial}{\partial t}\left(\left(\frac{\partial}{\partial t}+\frac{1}{T_2}\right)\right)+\frac{1}{T_1}\left(\frac{\partial}{\partial t}+\frac{1}{T_2}+\gamma^2 B_1^2 T_1\right)\right]\boldsymbol{M}_y=\frac{\gamma B_1 \boldsymbol{M}_o}{\boldsymbol{T}_1}$$

Solution
The code below was used to solve the above equation

```
    #include <iostream>
#include <fstream>
using namespace std;

double y01;
double y11;
double y=267513;
double B1=0.001;
double M=1;
ofstream out_data("/Users/emetere/Desktop/dee.txt");
void step(double dt)
{
for (double T1=0.75;T1<1.5;T1++)
  {
for (double T2=0.025;T2<0.2;T2++)
    {
      double a= (1/T1) +(1/T2);
double b= 1/(T1*T2)+ (y*y*B1*B1);
      double c=y*B1*M/T1;
double y2 = (2-a*dt-b*dt)*y11+(a*dt-1)*y01+(dt*dt*c);

  y01 = y11;
  y11= y2;
  }
  }
}

int de()
{
  y01 = 0;
  y11 = .1;
for(int i=0; i<100; i++)
  {

step(0.01*0.01);
    out_data << y11 << " "<<i<<endl;
```

```
  }
     const char *dirname="/Users/emetere/Desktop/dee.txt";
auto *tree = new TTree();
     TString dir = gSystem->UnixPathName(gInterpreter-
>GetCurrentMacroName());
      Double_t nlines = tree->ReadFile(Form(dirname,dir.Data()),"a:b");
  gROOT->SetStyle("Plain");
gStyle->SetOptStat(1111);
gStyle->SetOptFit(1111);
     TCanvas *c = new TCanvas("c", "General Plots",800,600);
tree->Draw("a:b");
  TGraph *gr = new TGraph(tree->GetSelectedRows(),tree->GetV2(), tree-
>GetV1());

     gr->Draw("AC");
  //gr->GetXaxis()->SetRangeUser(0.025, 0.2);
     // gPad->DrawFrame(xmin,ymin,xmax,ymax);

     //gPad->DrawFrame(0.025, -1, 0.2, 1);

  TH1 *frame = c->DrawFrame(0.025, -1, 0.2, 1);
     gr->Draw("same");
  frame->GetXaxis()->SetTitle("T2(s)");
  frame->GetYaxis()->SetTitle("My");
     frame->SetTitle("My @ Tau=10ms");
     gr->SetLineColor(4);
frame->SetMarkerStyle(15);
frame->SetMarkerSize(9);
     gr->SetLineWidth(6);
     gPad->Update();
     gPad->Modified();
c->Update();
c->Modified();
}
```

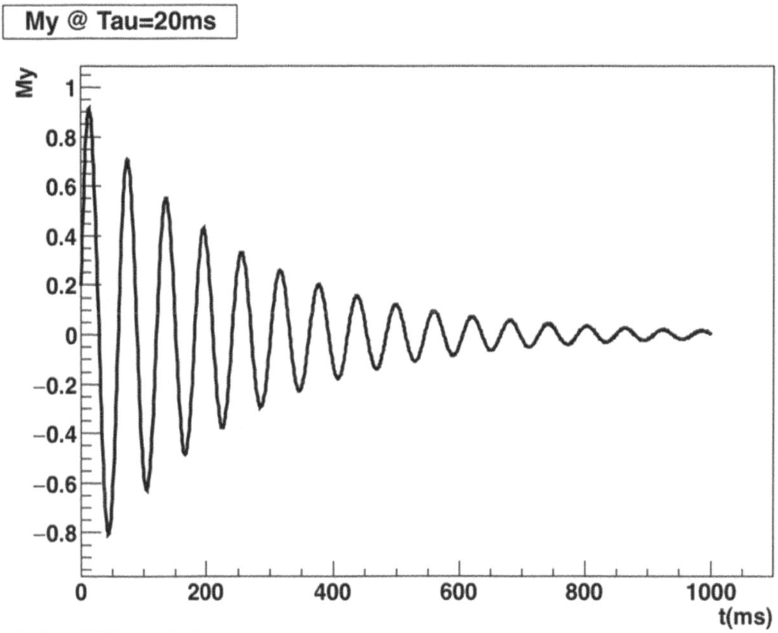
My @ Tau=20ms
My
1
0.8
0.6
0.4
0.2
0
−0.2
−0.4
−0.6
−0.8
0
200
400
600
800
1000
t(ms)

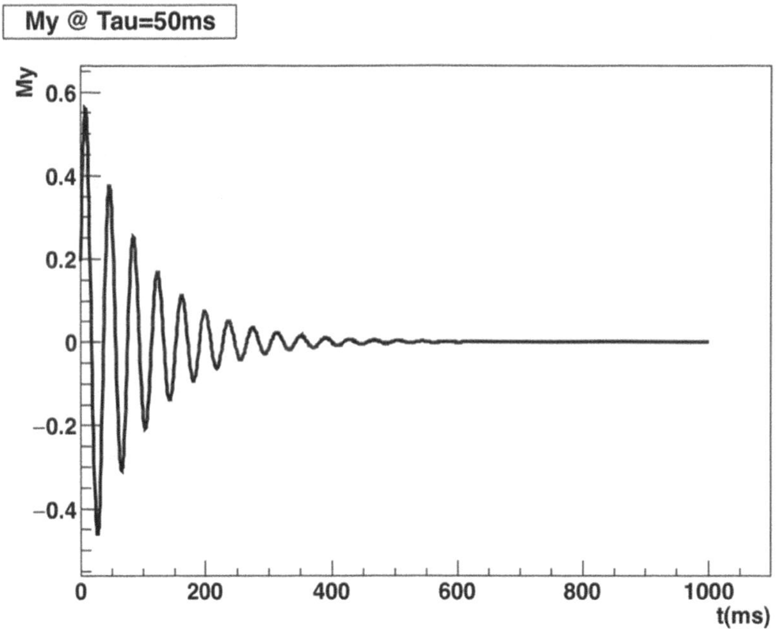
My @ Tau=50ms
My
0.6
0.4
0.2
0
−0.2
−0.4
0
200
400
600
800
1000
t(ms)

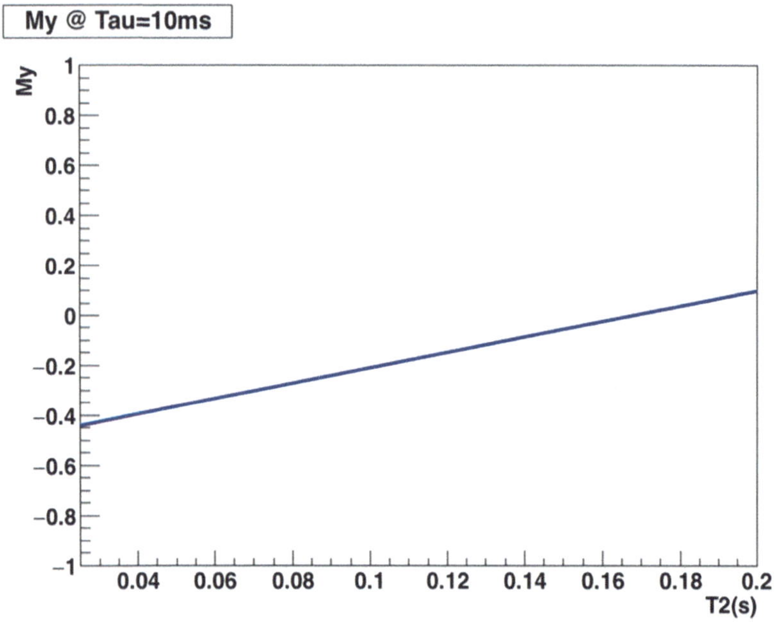

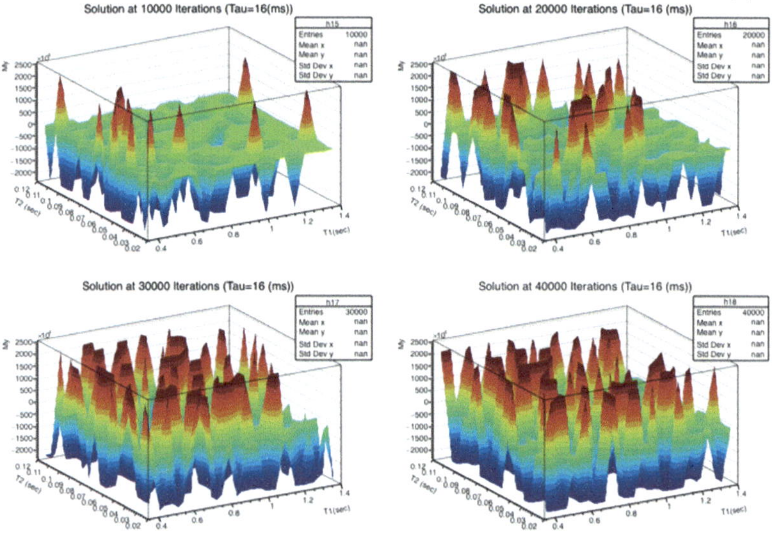

2.4 Numerical Analysis in Environmental Studies/Models

In this section various numerical analysis in environmental studies were discussed for readers to appreciate the wide application of numerical studies. In recent time, environmental scientists and professional choose software that directly relates to the kind of their analysis. Hence, it is difficult to:

i. improve on the framework of the software;
ii. link the software with libraries that makes it more sophisticated;
iii. perform software update without software developer;
iv. understand the mathematical framework of the software;
v. comprehend the scope of software efficiency.

In recent times, numerical modeling environment module is introduced to some software or simulators to enable software-users compare two simulations numerically. This feat has tremendously advanced environmental research with high precision. Also, mathematically inclined software-user can understand the accuracy of the model based on standard values.

Karacan et al. [3] worked on advances in the application of the most widely used grid-based numerical techniques, computational fluid dynamics (CFD) and numerical reservoir modeling, to assess gas control capabilities in longwall mines. Grid-based numerical modeling techniques are "non- classical" alternatives used in mining environment as a volume rather than as one-dimensional ventilation "network" branches. Different mining-related geometries with varying transport properties can be created within the simulation volume without developing mathematical models. The authors relied on proven works of National Institute for Occupational Safety and Health (NIOSH) in the U.S. and the Commonwealth Scientific and Industrial Research Organization (CSIRO) in Australia. The authors were able to show that CFD and reservoir simulations can complement each other. However, authors were restricted to (based on the parameters that were selected to gob gas ventholes, spontaneous heating management, gob inertization strategies, and methane control in development mining.

Hardesty et al. [4] worked on the understanding of micro-plastic (<5 mm diameter) distribution and pathways in the marine environment using numerical model simulations. The authors reviewed numerical models and ocean databases [OSPAR's marine beach litter program (http://www.ospar.org), International Coastal Cleanup (ICC) (http://www.oceanconservancy.org) and NOAA's marine debris program (http://www.marinedebris.noaa.gov)]. The authors focused on the numerical modeling of total plastic budget because knowledge on plastic in the oceans is insufficient to accurately estimate the total plastic budget. This challenge has extended influence on measuring or estimating ocean (micro) plastic directly at scale. The work did not show plots or simulation.

One vital parameter required for developing or executing numerical models is database. Database is a collection of dataset in whatever format i.e. asci, Binary NetCDF, HDF5. The data types in such formats may be signed byte (8 bits), unsigned char (8 bits), signed short integer (16 bits), signed long integer (32 bits), single-precision floating point (32 bits), double-precision floating point (64 bits). There are broadly six types of database, that is, centralized database, operational database, end-user database, commercial database, personal database and distributed database. Environmental database maybe sourced from Ocean Models, Static, In situ, remotely sensed and Climatologies which covers the following research areas listed below.

i. atmosphere database is comprised of air quality, climate, surface temperature, precipitation, solar irradiation, wind, tropospheric pressure profiles, clouds etc. Known database in this regards can be found in: ISCCP Cloud Data (https://www.ncdc.noaa.gov/isccp), SSMI-SSMIS Hydrological Products (https://www.ncdc.noaa.gov/ssmi), NDSC observations (www.ndsc.ncep.noaa.gov), Automated Smoke Detection and Tracking Algorithm (www.ospo.noaa.gov/Products/atmosphere/aerosol.html), HURSAT (www.ncdc.noaa.gov/hursat/), Vegetation Whole Globe Products (www.ospo.noaa.gov/Products/land/vegetation.html), MkIV observations (mark4sun.jpl.nasa.gov), WOUDC (www.woudc.org/index_e.html), MOZAIC observations (mozaic.aero.obs-mip.fr/web/), CARIBIC observations (www.caribic-atmospheric.com), ACE observations (www.ace.uwaterloo.ca), Aura observations (aura.gsfc.nasa.gov), ILAS observations (www.ilas.nies.go.jp/index.html), POAM observations (https://web.archive.org/web/20070301180400/ http://wvms.nrl.navy.mil/POAM/poam.html), NCAR-UCAR (https://rda.ucar.edu/), GIOVANNI (https://disc.gsfc.nasa.gov/datasets?Page=1), MISR (http://gdata1.sci.gsfc.nasa.gov/daac-bin/G3/download) etc.
ii. water database is comprised of quality/pollution, reserves, oceanic pressure etc. Some database that host information on water are: Daily OISST (www.ncdc.noaa.gov/sst/), Blended Sea Winds (www.ncdc.noaa.gov/data-access/marineocean-data/blended-global/blended-sea-winds), Validation Datasets—SURFA (www.ncdc.noaa.gov/oa/rsad/air-sea/surfa.html), Water quality data archive (http://environment.data.gov.uk/water-quality/index.html#), Spatial Data Catalogue (http://environment.data.gov.uk/ds/catalogue/#/catalogue) etc.
iii. terrestrial/soil database is comprised of land use, fertilization, salinization, farming, vegetative surfaces, forestry etc. Known database in this regards can be found in WISFAT (https://www.itu.int/en/ties-services/Pages/default.aspx), Crustal Structural Database Site (http://www.jamstec.go.jp/obsmcs_db/e/index.html), JAMSTEC Ocean-bottom Seismology Database (https://join-web.jamstec.go.jp/join-portal/en/), Global Seismic Travel time Database (http://d-earth.jamstec.go.jp/Traveltime/),European Soil Database (https://esdac.jrc.ec.europa.eu/content/european-soil-database-v20-vector-and-attribute-data), Harmonized World Soil Database (http://www.fao.org/soils-portal/soil-survey/soil-maps-and-databases/harmonized-world-soil-database-v12/en/), Soil Geographic Databases (https://www.isric.online/explore/soil-geographic-databases) etc.

Local database in this text refers to the database the user has direct control over its generation processes while external database refers to database that the users have no control over its generation processes (secondary users).Most environmental scientists and professionals are restricted to the usage of external database. In other words, they cannot know (in numeric terms) the error margin in the analysis. Most numerical analysis work seeks to merge theoretical models and environmental database to generate dataset that are only within theoretical limit. For example, ECMWF periodically uses its forecast models and data assimilation systems to 'reanalyse' archived observations, creating global data sets describing the recent history of the atmosphere, land surface, and oceans. It is based on this idea; a reanalysis process is illustrated.

i. **Step one**: understand the trend of the dataset. Dataset trend is very important because it is one of the means of detecting errors in external database. Also, data trend displays the dynamism of the graph for probing purposes. The atmospheric aerosol data was considered for this illustration. In understanding the trend of the dataset, the comparative analysis of the ground and satellite dataset is shown in Fig. 2.2. It is observed that the ground and satellite dataset have varying data trend. While satellite measurement has its highest value in June, the ground measurement has the lowest value in June. In such case, the synchronization of both measurements is very difficult. The reason for this varying measurements is the presence of moisture or cloud formation, precipitation rate [5].This scenario can also be statistically analyzed to see regression and deviation between both measurements.

ii. **Step two**: define theoretical limits of the forecast model. In this case, an emerging regional aerosol model [West African regional scale dispersion model (WASDM)] was used [6]. The relation between the mass of aerosol deposited, wind speed and aerosol content dynamics are shown in Fig. 2.3. The alignment of theoretical calculation and the dataset can be monitored using basic statistical tools.

iii. **Step three**: Characterize the different scenarios that makes up the graph. In this case, the aerosol optical depth dataset was collected with respect to four wavelengths i.e. 440, 555, 675 and 865 nm. First, the measurements were characterized by the aerosol size distribution (Fig. 2.4). Then the multiple computational characterization was used as shown in Figs. 2.4, 2.5, 2.6, 2.7, 2.8 and 2.9. In Fig. 2.5, the plots of the four wavelengths are shown. This helps to understand

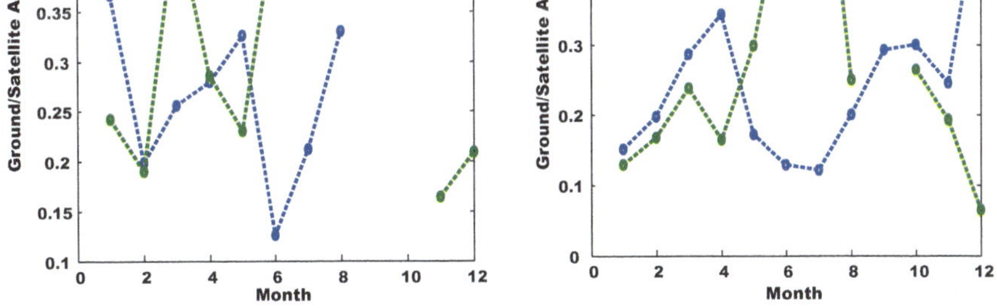

Fig. 2.2 Comparative analysis of Ground-Satellite measurement

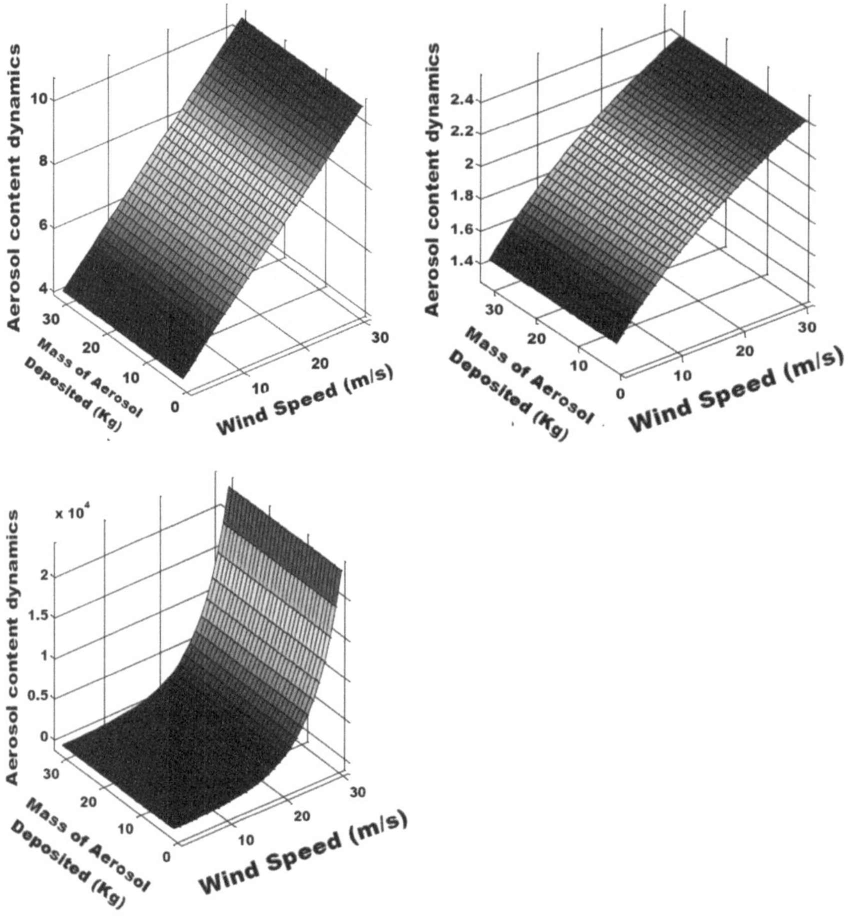

Fig. 2.3 Mass/wind relation in aerosol deposition

the variation between the AOD measurements based on their individual wavelengths. The comparative plots (2D, 3D and scattered plots) of the wavelengths are shown in Figs. 2.6, 2.7, 2.8 and 2.9.

iv. **Step five**: applying the forecast model to the database. In this case, WASDM was to numerically fit the dataset obtained from MISR. Dataset that are along the fits are extracted as reanalysis data (Fig. 2.10).

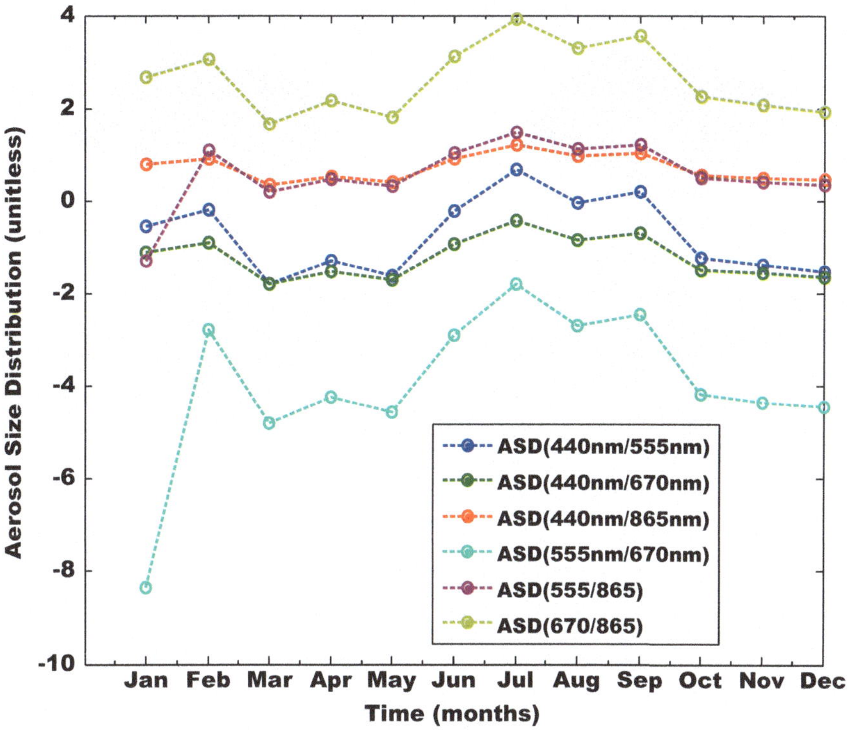

Fig. 2.4 Characterization using aerosol size distribution

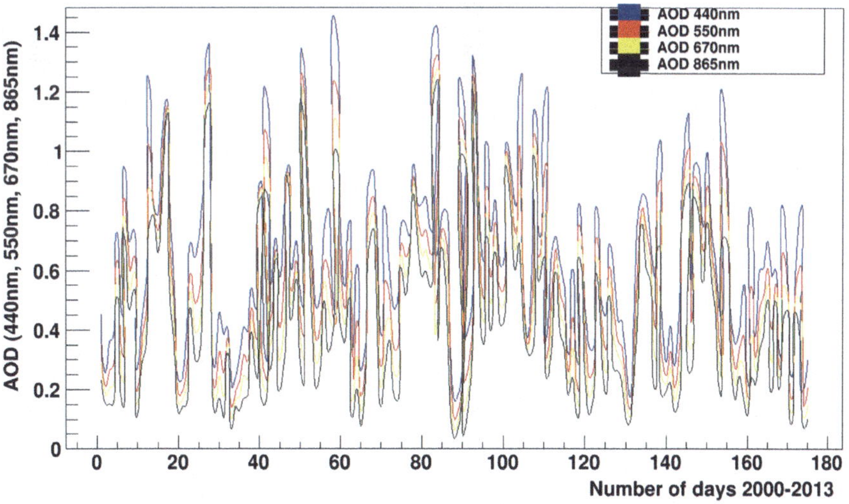

Fig. 2.5 AOD measurements via individual wavelengths

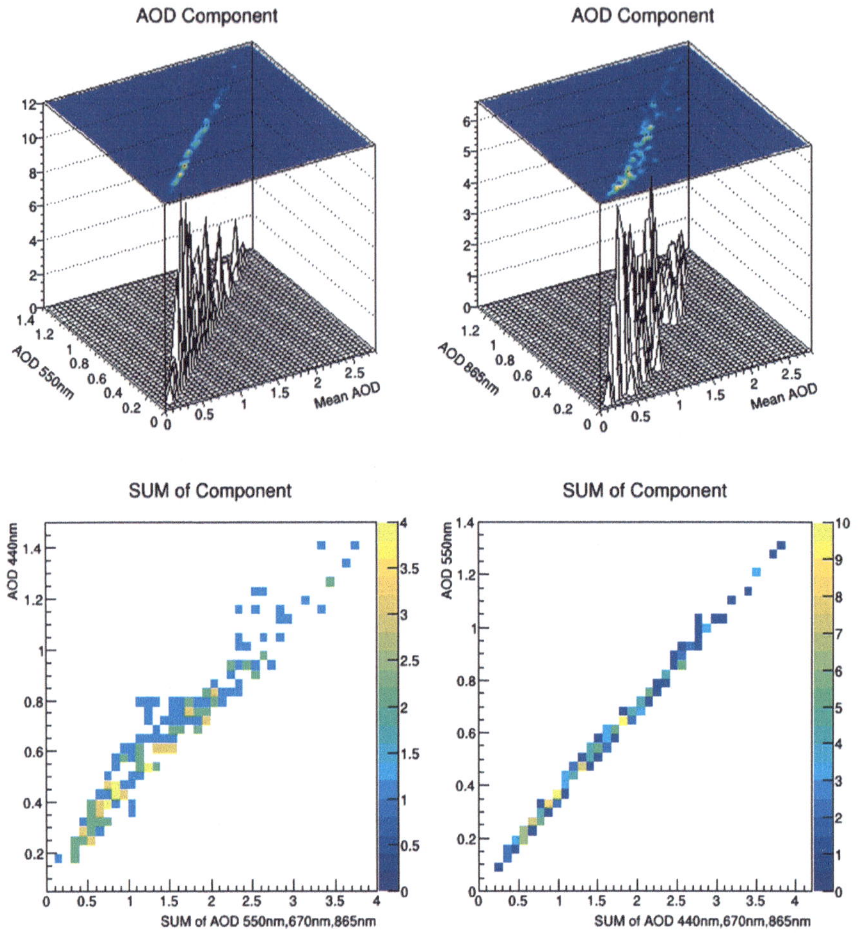

Fig. 2.6 Probing AOD measurement at 550 and 865 nm

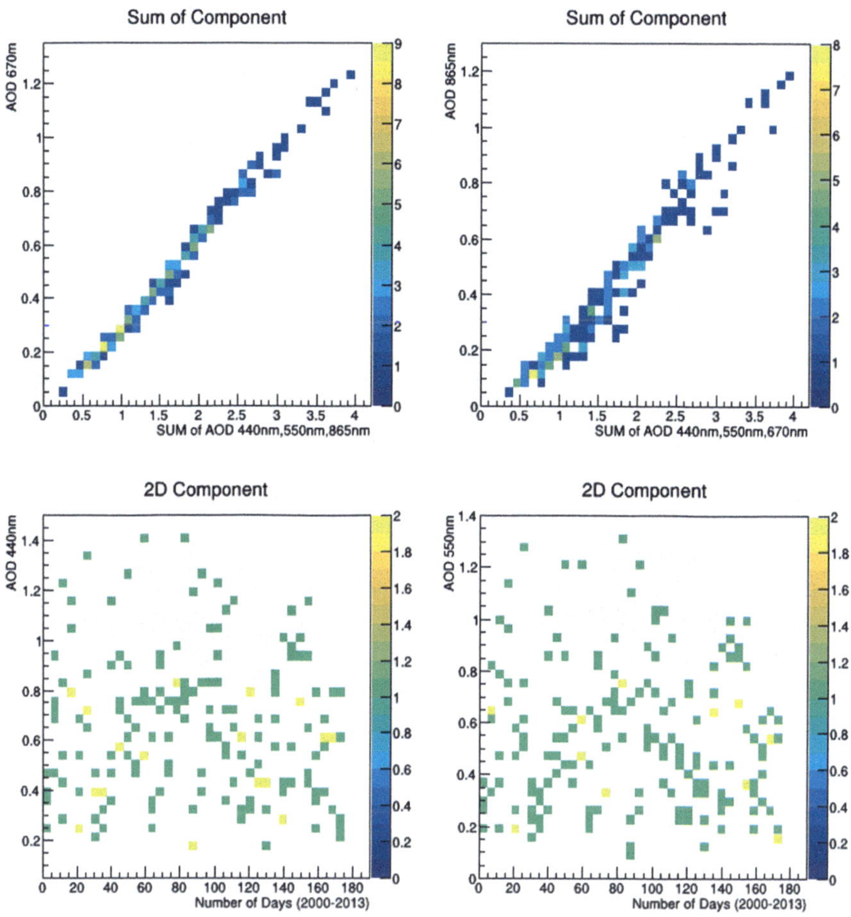

Fig. 2.7 Probing AOD measurement at 670 and 865 nm

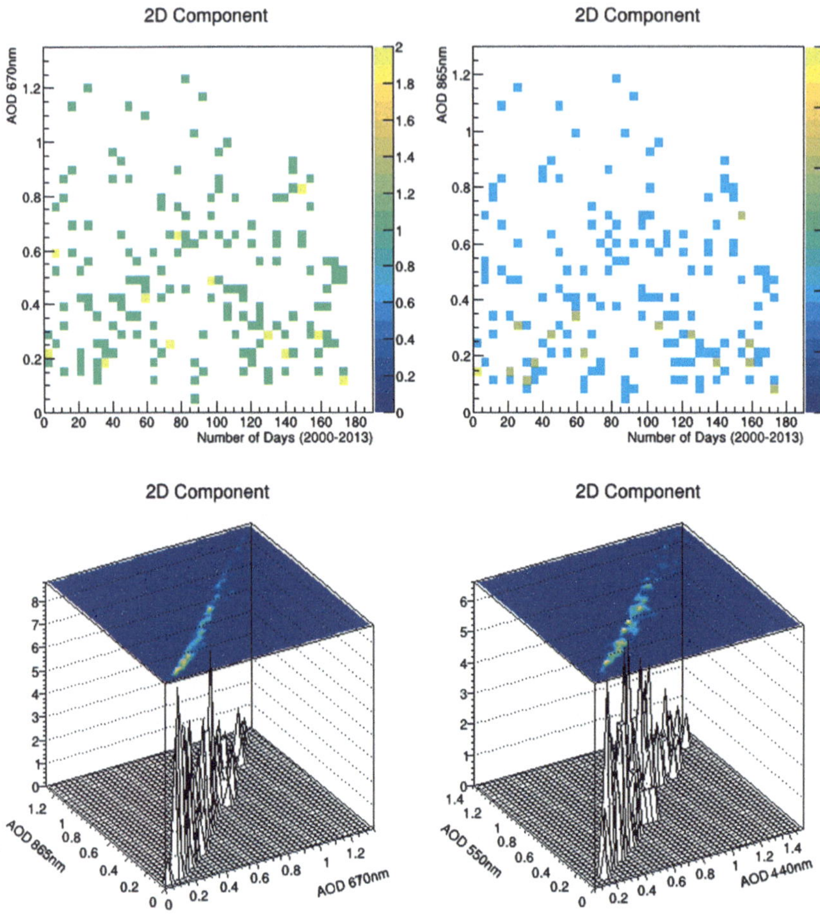

Fig. 2.8 Probing AOD measurement at 550 and 440 nm

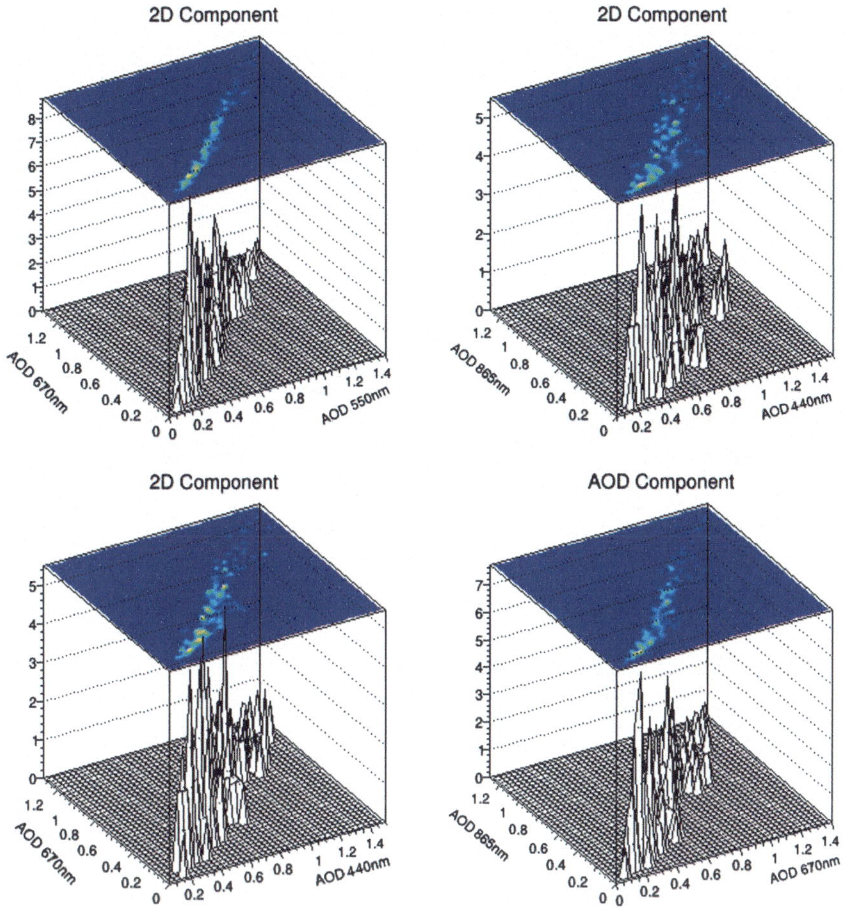

Fig. 2.9 Probing AOD measurement at 670 and 440 nm

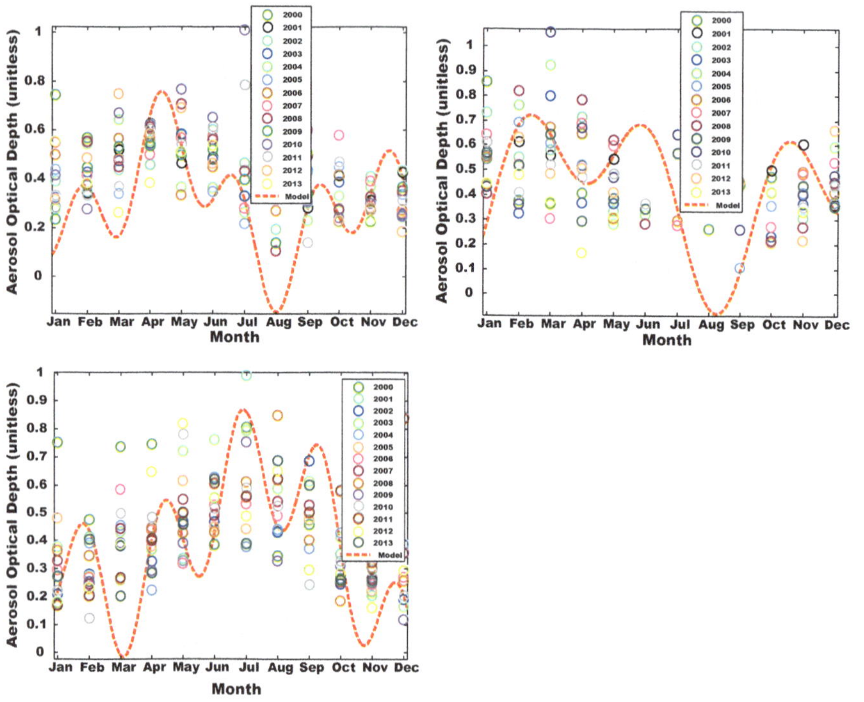

Fig. 2.10 Application of dataset to numerical model

References

1. Bates TS, Anderson TL, Baynard T, Bond T, Boucher O, Carmichael G, Clarke A, Erlick C, Guo H, Horowitz L, Howell S, Kulkarni S, Maring H, McComiskey A, Middlebrook A, Noone K, O'Dowd CD, Ogren J, Penner J, Quinn PK, Ravishankara AR, Savoie DL, Schwartz SE, Shinozuka Y, Tang Y, Weber RJ, Wu Y (2006) Aerosol direct radiative effects over the northwest Atlantic, northwest Pacific, and North Indian Oceans: estimates based on in-situ chemical and optical measurements and chemical transport modeling. Atmos Chem Phys 6:1657–1732
2. Thoma M (2013) Solving linear equations with Gaussian elimination. https://martin-thoma.com/solving-linear-equations-with-gaussian-elimination/. Accessed 24 July 2018
3. Karacan CÖ, Ren T, Balusu R (2006) Advances in grid-based numerical modeling techniques for improving gas management in coal mines. In: 12th U.S./North American mine ventilation symposium
4. Hardesty BD, Harari J, Isobe A, Lebreton L, Maximenko N, Potemra J, van Sebille E, Vethaak AD, Wilcox C (2017) Using numerical model simulations to improve the understanding of micro-plastic distribution and pathways in the marine environment. Front Mar Sci 4:30
5. Emetere ME (2016a) Generation of atmospheric constants over some locations in West Africa: a theoretical aid for measuring instruments design. Int J Eng Res Africa 27:119–146
6. Emetere ME (2016b) Numerical modeling of west Africa regional scale aerosol dispersion. A doctoral thesis submitted to Covenant University, Nigeria, pp 1–289

Chapter 3
Introduction to Computational Techniques

A computer is an integrated device that works on defined instruction to execute sequences of arithmetic or logical operations. The defined instruction is known as computer programming. Computer programming is the process of developing a set instruction in an executable computer program for executing specified operations. A good programme must be reliable (free of errors), robust (able to incorporate all variables), useable (easy to comprehend), portable (no complex components), maintainable (easy to update) and efficient (able to accomplish task). Software is an organized computer instructions or programs that dictate the operation of computer and its physical hardware from which the system is built. Software in a broad sense refers to application software (used by end or work users) and system software (i.e. operating systems and programs that execute application software). The general classification of application is namely productivity software (word processors, spreadsheets), presentation software, graphics software CAD/CAM software, specialized scientific applications, and industry-specific software. Software differs from computer program because software entails two or more programs that are carefully written to complement each other. The steps for developing software are illustrated in Fig. 3.1.

Another word that is commonly used in computer and will be discussed in succeeding chapter is the 'source codes'. Source code is any collection of customized instructions that is written in readable programming language, usually as plain text to facilitate computational task. An example of a source code is presented below in italics.

```
</style></head><body><div class="content"><pre class="codeinput">filename=<span class="string">'aerosol.xlsx'</span>;
sheet=1;
a=<span class="string">'A1:A9'</span>;
b=<span class="string">'B1:B9'</span>;
c=<span class="string">'C1:C9'</span>;
d=<span class="string">'D1:D9'</span>;
```


M. E. Emetere and E. T. Akinlabi, *Introduction to Environmental Data Analysis and Modeling*, Lecture Notes in Networks and Systems 58,
https://doi.org/10.1007/978-3-030-36207-2_3

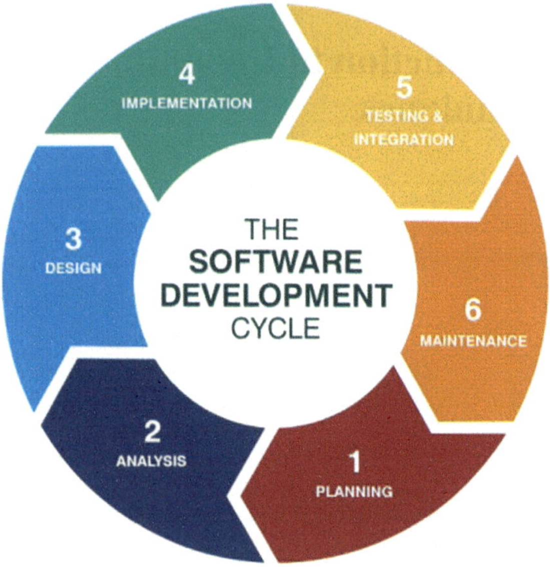

Fig. 3.1 Steps for software development

```
e=<span class="string">'F1:F13'</span>;
f=<span class="string">'G1:G13'</span>;
g=<span class="string">'H1:H13'</span>;
h=<span class="string">'I1:I13'</span>;
i=<span class="string">'K1:K6'</span>;
j=<span class="string">'L1:L6'</span>;
k=<span class="string">'M1:M6'</span>;
l=<span class="string">'N1:N6'</span>;
m=xlsread(filename,1,a);
n=xlsread(filename,1,b);
o=xlsread(filename,1,c);
p=xlsread(filename,1,d);
q=xlsread(filename,1,e);
r=xlsread(filename,1,f);
s=xlsread(filename,1,g);
u=xlsread(filename,1,h);
v=xlsread(filename,1,i);
w=xlsread(filename,1,j);
x=xlsread(filename,1,k);
y=xlsread(filename,1,l);
z=0.003;
A=0.076;
B=0.3;
```

```
C=2.73;
D=7.57;
E=14.8;
F=30.3;
E=((C-1)./m).^1;
F=((C-1)./n).^1;
G=((C-1)./o).^1;
H=((C-1)./p).^1;
I=(C-1)./q;
J=(C-1)./r;
K=(C-1)./s;
L=(C-1)./u;
M=(C-1)./v;
N=(C-1)./w;
O=(C-1)./x;
P=(C-1)./y;
Q=[1,2,3,4,5,6,7,8];
R=[1,2,3,4,5,6,7,8,9,10,11,12];
S=[1,2,3,4,5];
plot(Q,E,Q,F,Q,G,Q,H);
xlabel(<span class="string">'Month'</span>);ylabel(<span class="string">'Rate of mass transfer (unitless)'</span>);
legend(<span class="string">Wavelength(442 nm)'</span>,<span class="string">'Wavelength(555 nm)'</span>,<span class="string">'Wavelength(670 nm)'</span>,<span class="string">'Wavelength(865 nm)'</span>);
</pre><img vspace="5" hspace="5" src="aerosolmass_01.png" alt=""><p class="footer"><br>
Published with MATLAB&reg; 7.9<br></p></div><!--
```

A programming language is a formal language that highlights sets of instructions for execution. programming language is grouped by types namely Array languages, Assembly languages, Authoring languages, Constraint programming languages, Command line interface languages, Compiled languages, Concurrent languages, Curly-bracket languages, Dataflow languages, Data-oriented languages, Decision table languages, Declarative languages, and Embeddable languages. Array language is programming language that is used convert an operations on scalars to vectors, matrices, and higher-dimensional arrays. A typical example of modern languages that supports array language includes the following: Fortran 90, Mata, MATLAB, Analytica, TK Solver (as lists), Octave, R, Cilk Plus, Julia, Perl Data Language (PDL) and the NumPy extension to Python. An assembly language is a group of low-level programming languages used by microprocessors and other programmable devices to implement symbolic representation of machine code needed to program a given

CPU architecture. The type of assembly language includes: Complex Instruction-Set Computer (CISC), Reduced Instruction-Set Computer (RISC), Digital Signal Processor (DSP) and Very Long Instruction Word (VLIW).

Authoring language is a notation used to control the appearance and functionality of WebPages when displayed in a browser. Example of authoring language are DocBook, DITA, PILOT, TUTOR, Bigwig, Chamilo, Hollywood (Hollywood Designer graphical interface), Learning management system, SCORM, Web design program, XML editor and Game engine. Constraint programming is a type of programming paradigm that displays variable in the form of constraints. Libraries that accepts constraint programming are Artelys Kalis, C++, Java, Python library, FICO Xpress module, Cassowary, Smalltalk, Ruby library (LGPL), CHIP V5, Choco, Java library, Comet, Cream, Java library (LGPL), Disolver, Facile, OCaml library (CC0 1.0), finite-domain, Haskell library (MIT), Gecode, Google OR-Tools, JaCoP, Java library, LINDO MonadicCP, Haskell library (BSD-3-Clause) etc. Command line interface languages are languages that are used to interact with a computer program where the user issues commands to the program in the form of successive lines of text. The successive lines of text are called command lines. The command lines are executed in the shell of operating system. The shell of operating system includes AmigaOS (Amiga CLI/Amiga Shell), Unix OS (Bourne shell, Almquist shell, Debian Almquist shell, Bash, Korn shell, Z shell, C shell, TENEX C, Emacs shell, rc shell rc, Stand-alone shell and Remote shell), Microsoft Windows (CMD.EXE, Windows PowerShell, Hamilton C shell, 4NT, Recovery Console), DOS(COMMAND.COM, 4DOS, NDOS and GW-BASIC), OS/2 (CMD.EXE, Hamilton C, and 4OS2), IBM OS/400 (AS/400 Control Language, iSeries QSHELL) Apple (Apple DOS/Apple ProDOS) and Mobile devices (DROS, Java ME platform).

Compiled languages is a programming language whose implementations are typically compilers (translators that generate machine code from source code), and not interpreters (step-by-step executors of source code). Examples of compiled programming language are ALGOL, BASIC, C, D, CLEO, COBOL, Cobra, Crystal, Eiffel, Fortran, Go, Haskell, Haxe, JOVIAL, Julia, LabVIEW, G, Pascal, SPITBOL, Visual Foxpro and Visual Prolog. Embeddable languages are programming language that supports scripting in realtime systems. Example of embedded language includes Python/C++/Java compiler.

Library is a collection of non-volatile resources and implementations of behaviour that is written in a specific modern language to invoke operations in a well-defined interface. A computer library includes configuration data, documentation, help data, message templates, pre-written code and subroutines, classes, values or type specifications. Libraries links or binds computational processes using a linker or binder program that searches a set of libraries or modules to be executed. The loader program is used to create links to libraries by adjusting its references. When linking is performed during the creation of an executable or another object file, it is known as static linking or early binding. This kind of library is known as the static library. In the static library the linking process is accomplished using a linker or compiler. A typical function of a linker is shown in Fig. 3.2. Also, a typical function of compiler is shown in Fig. 3.3. When a file is shared by executable files and object files, it is

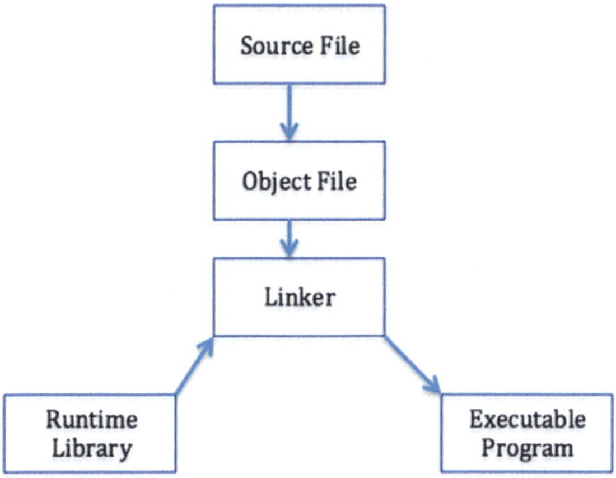

Fig. 3.2 Function of a linker

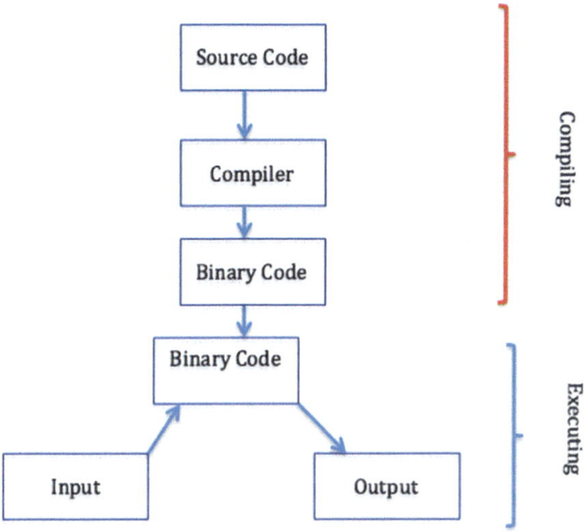

Fig. 3.3 Function of a compiler

referred to as a shared library. Shared objects (that has been statically or dynamically linked) moves to the memory at load time or run time, rather than being copied by a linker when it creates a single monolithic executable file for the program. Dynamic linking is a computational linking performed while a program is being loaded (load time) or executed (run time), rather than when the executable file is created. A sub-set of dynamic linking is dynamic loading which entails a dynamically linked library loading and unloading at run time on request.

3.1 General Outline of Computational Techniques

What generally comes to the mind when computational technique is mentioned is computer software. Computer software is structured algorithms or codes that implement a specific task. Hence, the three broad classification of computer software are, namely, system software, application software and programming software. System software is made-up of operating system (Microsoft Windows, Mac OS X, and Linux) device drivers (Bios, motherboard drivers, hardware divers, virtual device drivers, sound cards etc.), servers and software components. Application software (AS) is used for attaining specific tasks. The types of AS are: licensed, sold, freeware, shareware, and open source. In this chapter, we shall be emphasizing on the open source since most environmental models are open source. Programming software are mini-codes or macros used for writing programs through tools such as editors, linkers, debuggers, compilers/interpreters etc.

Computational techniques may be seen as a very broad concept in modern research. The early type of computational technique is termed computer aided algebraic systems. In this technique, algebraic related problems are solved using algorithms, codes, charts and special syntaxes. However, as research expanded, there was need for more application software for special task. In environmental modeling, there is the general and specific application software. The general application software performs tasks as speculated by the modeller discretion. Example of general application software in environmental modeling includes MATLAB, MATHCAD, Origin, Homer, GNU plot, FreeMat, Octave, Microsoft Excel, Magma, Maple, R package, Mathematical, SageMath, PolyMath, SMath, COMSOL etc. The various computational techniques are shown pictorially in Figs. 3.4 and 3.5.

The outline of the computational technique will be explained using the pictorial illustration in Fig. 3.6.

The fabrication stage is referred to the developmental stages where the modeller decides on what kind of computational technique would be appropriate for

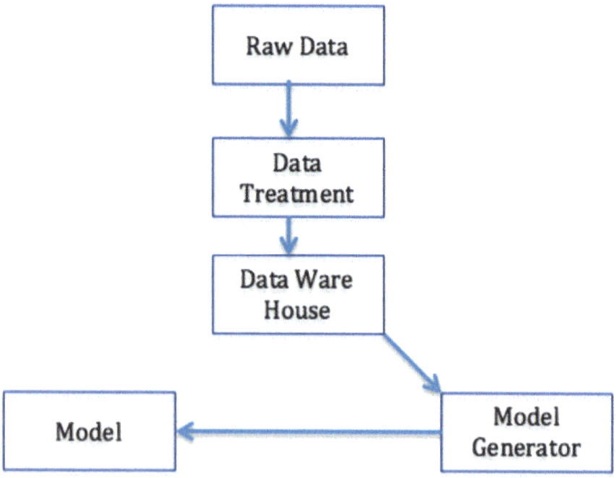

Fig. 3.4 Computational technique for data mining

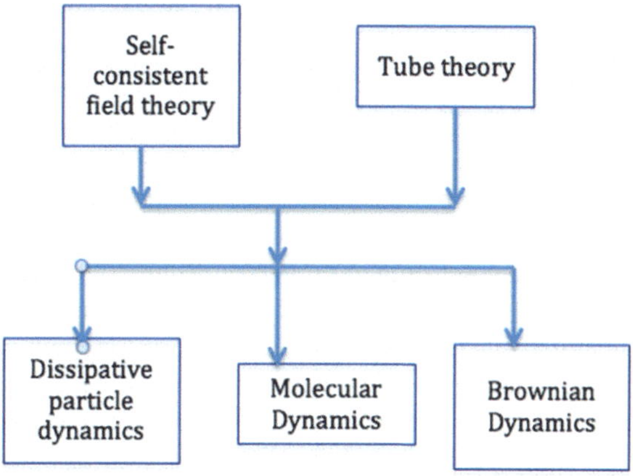

Fig. 3.5 Computational technique for materials

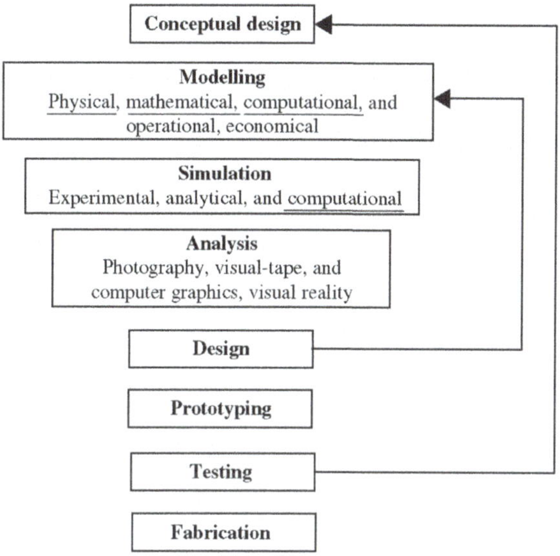

Fig. 3.6 pictorial illustration of computational technique

the model. Most modeller at this stage decides to adopt mathematical methods. Mathematical method is a tool for solving abstract and real problems. Mathematical method is fondly used in science, engineering, social science and humanities. The list of method that are usually considered in mathematical methods includes: Infinite series, power series, Complex numbers, Integral transform, Wavelet transform, Fourier transform spectroscopy, Harmonic analysis, Linear algebra, Partial differentiation, Multiple integrals, Vector analysis, Fourier series and transforms, Ordinary differential equations, Calculus of variations, Linear function, Quadratic function, Cubic function, Quartic function, Cabibbo-Kobayashi-Maskawa matrix, Density matrix, Fundamental matrix, Fuzzy associative matrix, Gamma matrices,

Gell-Mann, Hamiltonian matrix, Wall polynomial, Wangerein functions, Weber function, Weierstrass function, Weisner's method, Whittaker function, Wilson polynomial, Irregular matrix, Overlap matrix, S matrix, State transition matrix, Substitution matrix, Z-matrix, Quintic function, Sextic function, Tensor analysis, Special functions, Schubert polynomial, Schur polynomial, Selberg integral, Sheffer polynomial, Slater's identities, Stieltjes polynomial, Stieltjes–Wigert polynomials, Strömgren function, Struve function, Legendre function, Bessel function, Hermite function, Laguerre function, Partial differential equations, Functions of a complex variable, Integral transforms, Gamma function, Barnes G-function, Beta function, Digamma function, Polygamma function, Incomplete beta function, Incomplete gamma function, K-function, Multivariate gamma function, Student's t-distribution, Probability and statistics, Two-sided Laplace transform, Mellin transform, Laplace transform, Fourier transform, Fourier series, Sine and cosine transforms, Hartley transform, Short-time Fourier transform, Celine's polynomial, Charlier polynomial, Pafnuty Chebyshev, Chebyshev polynomials, Painlevé function, Painlevé transcendents, Poisson–Charlier polynomial, Pollaczek polynomial, Cyclotomic polynomials, Rectangular mask short-time Fourier transform, Gegenbauer polynomials, Gottlieb polynomial, Gould polynomial, Gudermannian function, Chirplet transform, Fractional Fourier transform (FRFT), Hankel transform, Hall polynomial, Hall–Littlewood polynomial, Hankel function, Heine functions, Racah polynomial, Riccati–Bessel function, Riemann, zeta-function, Rodrigues formula, Rogers–Askey–Ismail polynomial, Rogers–Ramanujan identity, Rogers–Szegő polynomials, Hermite polynomials, Heun's equation, Horn hypergeometric series, Hurwitz zeta-function, Boubaker polynomial etc. However, some modellers would want to generate their own brand of mathematical methods.

Secondly, the modellers choose what kind of programming language would be appropriate to solve the mathematical method that has been adopted. The common programming language used in research in modern times include: Python, Q# (Microsoft programming language), C, C−−, Java, C++, C#, MATLAB, Mathcad, Visual Fortran Visual FoxPro, JavaScript, JCL, Jython, MATH-MATIC, Visual J++, Visual J#, Mathematica etc. Some modeller used one or more of the programming language mentioned above.

As shown in Fig. 3.6, the modeller tests the computationally solved problem on dataset to see the performance of the adopted solution. If the modeller is not satisfied with the outcome, he checks his computational procedures once more until he/she can figure-out a satisfactory outcome. Once the modeller is satisfied with the testing of the computational work, he moves on to the prototyping of the computational work. Most modeller stops at the testing stages because of funds or technical know-how. The prototype is sent to research centers or industries for feedback purposes. Once the feedback comes out from the sources, the modeller goes on to the designing stage. At this point the computational work comes out as a licensed or open-source application. The modeling, analysis and simulation stages are performed by the end-users.

The specific application software [1] for environmental modeling includes SimaPro or GaBi (for assessing Carbon Footprints), Qual2K (river and stream water quality and process interaction model), Hec-HMS (for hydrologic modeling system), WaterCAD (for designing and planning hydrologic systems), ADORA or AERMOD (for monitoring air pollution), EMEX (for managing incidents and track corrective and preventative actions), Enviro (for gathering, managing, and displaying lab and field data for water, soil, and air), SmartData (for investigating environmentally contaminated sites), Accuvio (for monitoring carbon print and greenhouse gases) etc. In recent time, it has been proven that some environmental application software has their flaws [2]. This development is quite worrisome because beginners and young professionals in environmental studies may be building on faulty foundation.

Based on the aforementioned, environmental scientists are advised to use either the general application software or adopt open-source application/library to prevent uncontrollable error-prone analysis/research.

3.2 Open Source Scientific Packages

The term open source in the context of software development refers to computer programs that are free to view, use, alter and share. Authors of open source application or library benefit from experts in the field to enhance the functionality of the application. In other words, open source is driven by the aim of product development. The list of open source application includes FreeSWITCH, Jitsi, QuteCom, Slrn, Twinkle, Ring, Tox, Geary, GNU Queue, HTCondor, OpenLava, Apache Axis2, Apache Geronimo, Bonita Open Solution, GlassFish, Jakarta Tomcat, JBoss Application Server, ObjectWeb JOnAS, TAO, Enduro/X, Akregator, Liferea, RSS Bandit, Sage, Popcorn Time, qBittorrent, Liferay, Oxwall, uPortal, FreeNX, OpenVPN, rdesktop, Synergy, VNC, Remmina, Edoceo Imperium, FrontAccounting, GnuCash, LedgerSMB, TurboCASH, OpenDisc, ASSP, MailScanner, Scrollout F1, SpamAssassin, SpamBayes, ClamAV, ClamTk, ClamWin Free Antivirus, Caffe, CNTK, Deeplearning4j, Distributed Machine Learning Toolkit, H2O, NuPIC, OpenCyc, OpenNN, SystemML, TensorFlow, Torch, Celestia, KStars, Stellarium, Amarok, Ardour, aTunes, Audacious, Audacity, CDex, Cdrtools, cdrtfe, Clementine, DeaDBeeF, EasyTAG, Exaile, FlacSquisher, Fre:ac, Frinika, Giada, Guayadeque, Hydrogen, Jajuk, Jams, KMid, Linux MultiMedia Studio, Mixxx, MOC, Mp3splt, MuseScore, Nightingale, orDrumbox, Qmmp, Radio Downloader, Rhythmbox, SoX, TEncoder, XiX Music Player, xwax, Yoshimi, ZynAddSubFX, AMANDA, Areca Backup, Attic, Backup, Backupninja, BackupPC, Back In Time, Bacula, Bareos, Box Backup, BURP, Clonezilla, Create Synchronicity, DAR, DirSync Pro, DriverBackup, Duplicity, FOG, FreeFileSync, FullSync, Grsync, LuckyBackup, Mondo Rescue, Obnam, Partimage, Redo, Rsnapshot, Rsync, SafeKeep, SMS Backup+, SnapBackup, Synkron, Unison, UrBackup, Weex, Win32DiskImager, XSIbackup, Alluxio, Ambari, Avro, Cascading, Chukwa, Data Torrent RTS, Disco, Flume, Genie, Hadoop, Hadoop Distributed File System, HPCC, Hypertable, Ignite, Kudu, Lipstick, Lucene, Lumify, MapReduce, Mesos, Oozie, Pandas, Pig, Pidgin, GanttPV, FreeMind, VirtualBox, Notepad++, Greenshot, FileZilla, TurboCASH 5, GnuCash,

Ares Galaxy, eMule, Tribler, GIMP, Blender 3D, Dia, Inkscape, Scribus, VLC Media Player, SMPlayer, Handbrake, Audacity, Media Player Classic, KeePass Password Safe, TrueCrypt, Bitdefender Antivirus Free Edition, HealthMonitor, OpenX, Asterisk, Ekiga, FreePBX etc.

Unlike the open source, the source code of the "proprietary" or "closed source" software has exclusive control by its author. Open source licenses enable computer users' permission to use an application for any purpose. However, it is mandatory anyone who releases a modified open source program to release the source code for that program alongside it. There are times when developers focus completely on adding features without taking cognizance of code errors that creep in and render the software unusable. In recent times, code errors are exterminated from the software configuration management using a Continuous integration (CI) server. CI function as the gatekeeper that helps keeps a tab on code quality. The type of CI include Jenkins, Buildbot, Travis CI, Strider, Go and Integrity. Jenkins was developed using Java. Buildbot was developed using python. Strider was developed using Node.JS and JavaScript. Integrity was developed by Ruby.

Most authors of open source software have the advantage of building a better code, learn new developmental packages (from expert contributions), secure the software to their name, give back to their immediate and extended communities, and stand the opportunity to receive free tools from highly-placed organization. The challenge of the open source software is that it is most time tricky to compile or install. Most of the open source software do not have tutorial center or help website for debugging issues. Also, some of the open source software lack user-friendly interfaces and most time have compatibility issues (due to some specialized driver that must be installed before set-up). Most open source software has update issues especially if the authors are not funded to create awareness among end-users. Should there be an extended damage due to corrupt files, there is no liability or infringement indemnity protection. Some of the operating system where the open source software is used is difficult to learn especially if the user is used to a particular kind of operating system. Most of the latest hardware is incompatible to the open-source platform. Hence, user may rely on third-party drivers to aid the software compilation and feature performance.

Looper [3] gave a summary of how to build an open source application. The factor to be considered includes: select the platform to project your work; write chatbots and integrate them with Slack; make available the necessary tools that developers may need to improve on the application; seek helpers and maintain a good relationship with them; make use of blogs to harvest feedbacks you may need to improve upon the software.

Open source library uses some open source software like Linux, Apache Web Server, OpenOffice, GIMP, Audacity, and Firefox browsers. Koha and Evergreen are referred to as integrated library systems (ILS). They are the most popular libraries and both are licensed under a GNU General Public license. The pictorial illustration on how to build open source application is shown Figs. 3.7 and 3.8.

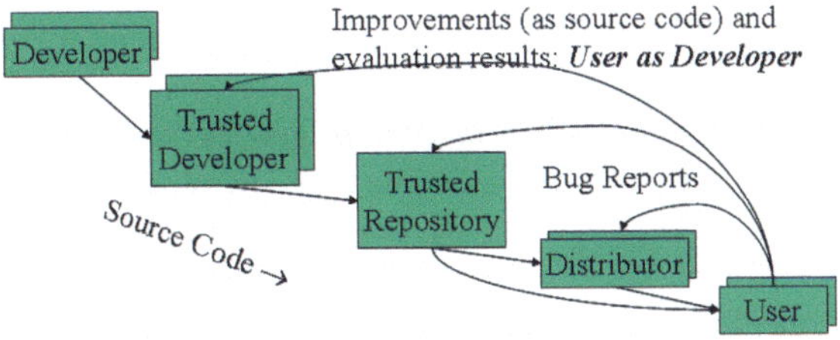

Fig. 3.7 Building open source applications: users as developer [4]

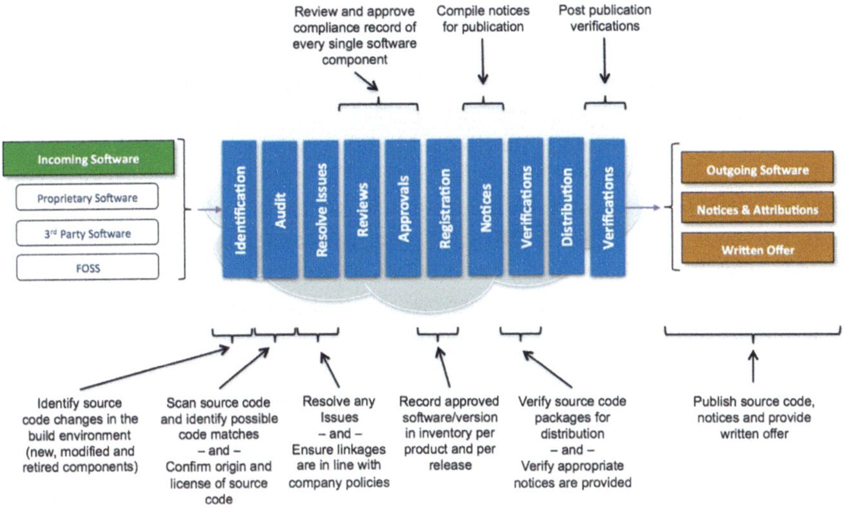

Fig. 3.8 Building open source application [5]

3.3 Open-Source Library

library is a collection of non-volatile resources used by computer programs, often to develop software. Libraries are used for software development to enhance the software to perform specific task. The major computer libraries are written in terms of language e.g. Multi-language, C, C++, Delphi, .NET Framework languages (C#, F#, VB.NET and PowerShell), Fortran, Java, Scala, Perl, Python, Groovy, XNUMBERS, INTLAB etc.

The C++ libraries includes: Boost, GSL, BDE, Dlib, JUCE, Loki, Reason, yomm2, Folly, Abseil, cxxomfort, libsourcey, OnPosix, Ultimate++, CAF, cpp-mmf, CommonPP, Better Enums, Smart Enum, nytl, SaferCPlusPlus, fcppt, bitfield.h, composite_op.h, Yato, Kangaru, yaal, rpnx-serial, libnavajo, C++ RESTful framework, C++ REST SDK, cpr, cpp-netlib, cpp-redis, tacopie, Boost.Asio, Boost.Beast, gsoap, POCO, omniORB, ACE, TAO, wvstreams, Unicomm, restful_mapper, zeromq, curlpp, Apache Thrift, libashttp, Simple C++REST library, libtins, PcapPlusPlus,

HTTPP, The Silicon C++14 Web Framework, ngrest, restc-cpp, OpenDDS, Breep, uvw, rest_rpc, EasyHttp, nghttp2, Dear ImGui, FLTK, nana[, WxWidgets, OWL-Next, tiny file dialogs, Switch, glibmm, gtkmm, goocanvasmm, libglademm, libgnomecanvasmm, webkitgtk, flowcanvas, evince, Qt, qwtplot3d, qwt5, libdbusmenu-qt, QuickQanava, QuickProperties, SFML (Simple and Fast Multimedia Library), SDL (Simple DirectMedia Layer), SIGIL (Sound, Input, and Graphics Integration Library), Cinder, openFrameworks, cairomm, nux, pangomm, gegl, stb, Adobe/boost GIL, GraphicsMagick, Skia Graphics Engine, enwiki:Skia_Graphics_Engine, Anti-Grain Evolution, plotutils, libraw, openexr, qimageblitz, imagemagick, djvulibre, poppler, SVG++, id3lib, taglib, opencv, dlib, ITK, OTB, Vulkan, OpenGL, bgfx, Ogre3D, Diligent Engine, GLEW, GLAD, Epoxy, GLFW, GLM, hlsl++, assimp, VTK, Magnum, Irrlicht, Horde3D, Visionaray, Open CASCADE, OpenScene-Graph, EntityX, Anax, EntityPlus, EnTT, BOX2D, stats++, StatsLib, alglib, Array-Fire High Performance Computation Library, GNU MP bignum C++ interface, BigNumber, Boost.Multiprecision, Boost.Math.Special Functions and Statistical-istributions, Boost.Random, NTL—A Library for doing Number Theory, cpp-measures, G+Smo cross-platform library for isogeometric analysis, Exact floating-point arithmetic library, Boost.uBLAS, Eigen, Armadillo, Blitz++, IT++, Dlib—linear algebra tools, Blaze, ETL, DecompLib, OptimLib, Boost.Graph, LEMON, OGDF—Open Graph Drawing Framework, NGraph—a simple (Network) Graph library in C++, GTpo, cln, Dlib—machine learning tools, MLPACK—machine learning package, Shogun—large scale machine learning toolbox, CGAL—Computational geometry algorithms library, Wykobi—Computational geometry library, PCL—Point Cloud library, yasmine—C++11 UML state machine framework, libxml++, pugixml, tinyxml, tinyxml2, Xerces, gSOAP, ai-xml, json, ArduinoJson, jsonme–, ThorsSerializer (JSON/YAML Input Output Streams), JsonBox, jsoncpp, zoolib, JOST, CAJUN, libjson, nosjob, rapidjson, jsoncons, JSON++, qjson, json-cpp, jansson, json11, JSON Voorhees, jeayeson, ujson, minijson, jios (JSON Input Output Streams), Botan, gnutls, openssl, crypto++, TomCrypt etc.

The Python libraries includes: ADOdb, AppJar, Beautiful Soup (HTML parser), CGAL, CheetahTemplate, Construct (python library), Cubes (OLAP server), Genshi (templating language), Gensim, IronPython, Jinja (tmplate engine), Kamaelia, Kid (templating language), Kivy (framework), Natural Language Toolkit, Pickle (Python), PLWM, PyEphem, Pygame, Pyglet, PyGObject, PyGTK, PyObjC, PyQt, PySide, Python Imaging Library, Python Robotics, Python-Ogre, RDFLib, Redland RDF Application Framework, Requests (software), RPyC, SimpleITK, SimPy, Sound Object (SndObj) Library, Soya3D, SpaCy, SQLAlchemy, SQLObject, Storm (software), Tkinter, Topsite Templating System, Twisted (software), VPython, WxPython, XDMF, Pipenv, PyTorch, Caffe2, Pendulum, Dash, PyFlux, Fire, imbalanced-learn, FlashText, Luminoth, Scrapy, Pillow, NumPy, SciPy, matplotlib, Scapy, pywin32, nltk, nose, SymPy, IPython etc.

Open source library uses some open source software like Linux, Apache Web Server, OpenOffice, GIMP, Audacity, and Firefox browsers. Koha and Evergreen are referred to as integrated library systems (ILS). They are the most popular libraries and both are licensed under a GNU General Public license. Opens source application depends on open source libraries to perform certain operation or features. Example of the dependent open source applications are as follows: OpenCog, OpenCV, TREX, ROS, YARP, FreeCAD, BIM, LibreCAD, Blender, flightgear, SimPy, Scribus, Bitcoin Core, Bonita Open Solution, CiviCRM, Compiere, Cyclos, Dolibarr, ERPNext, GnuCash, HomeBank, iDempiere, Ino erp, jFin, JFire, KMyMoney, LedgerSMB, metasfresh, Mifos, Odoo, Openbravo, OrangeHRM, Postbooks, QuickFIX, QuickFIX/J, SQL Ledger, SugarCRM, Tryton, TurboCASH, Wave Accounting, ZipBooks, Evergreen, Koha, NewGenLib, OpenBiblio, PMB, refbase, Darktable, digiKam, GIMP, Inkscape, Krita, LightZone, RawTherapee, Chemistry Development Kit, JOELib, OpenBabel, P-GRADE Portal, CellProfiler, Endrov, FIJI (software), Ilastik, ImageJ, IMOD, ITK, KNIME, VTK, 3DSlicer, Abalone, Ascalaph Designer, GROMACS, LAMMPS, MDynaMix, NAMD, NWChem, Avogadro, BALLView, Jmol, Molekel, MeshLab, PyMOL, QuteMol, RasMol, Ninithi, CP2K, LimeSurvey, CMU Sphinx, Emacspeak, ESpeak, Festival Speech Synthesis System, Modular Audio Recognition Framework, NonVisual Desktop Access, Text2Speech, Dasher, Gnopernicus Virtual Magnifying GlassEnvironment, Konstanz Information Miner (KNIME) OpenNN, Orange, RapidMiner, Scriptella ETL, Weka, JasperSoft, ParaView, VTK, ResourceSpace, ApexKB, Lucene, Nutch, Solr, Xapian, Elasticsearch, Konstanz Information Miner (KNIME), Pentaho, SpagoBI, Talend, OpenAFS, Tahoe-LAFS, CephFS, OpenX, Asterisk. Ekiga, FreePBX, FreeSWITCH, Jitsi, QuteCom, Enterprise Communications System sipXecs, Twinkle, Ring, Tox, Geary, Mozilla Thunderbird, GNU Queue, HTCondor OpenLava, pexec, Apache Axis2, Apache Geronimo, Bonita Open Solution, GlassFish, Jakarta Tomcat, JBoss Application Server, ObjectWeb JOnAS, TAO, Enduro/X, Akregator, Liferea, RSS Bandit, RSSOwl, Sage, Popcorn Time, qBittorrent, Drupal, Liferay, Oxwall, Sun Java System Portal Server, uPortal, FreeNX, OpenVPN, rdesktop, Synergy, VNC, Remmina, Brave, Chromium, Firefox, Midori, Tor Browser, Waterfox, SeaMonkey, Cheese, Guvcview, cURL, HTTrack, Wget, Apache Cocoon Apache, AWStats, BookmarkSync, Cherokee, curl-loader, FileZilla, Hiawatha, HTTP File Server, lighttpd, Lucee, Nginx, NetKernel, Qcodo, Squid, Vaadin, Varnish, XAMPP, Zope, ATutor, Chamilo, Claroline, DoceboLMS, eFront, FlightPath, GCompris, Gnaural, H5P, IUP Portfolio, ILIAS, Moodle, OLAT, Omeka, openSIS, Sakai Project, SWAD, Tux Paint, UberStudent, KGeography, Kiten KVerbos WINE, CyberBrau, Pencil2D, Pivot Animator, Synfig, Tupi (formerly KTooN), OpenToonz, Blender, OpenFX, Seamless3d, Pencil2D, SWFTools, Eye of GNOME, F-spot, Geeqie, Gthumb, Gwenview, Kphotoalbum, Opticks, Dr. DivX, FFmpeg, MEncoder, OggConvert, Avidemux, AviSynth, Blender, Cinelerra, DScaler, DVD Flick, Flowblade, Kaltura, Kdenlive, Kino, LiVES, Natron, OpenShot Video Editor, Pitivi, Shotcut, VirtualDub, VirtualDubMod, VideoLAN Movie Creator, Avidemux, HandBrake, FFmpeg, Apache OpenOffice, Calligra Suite, LibreOffice, Chandler, KAddressBook, Kontact, KOrganizer, Mozilla Calendar, Novell Evolution, OpenSync, Project.net, TeamLab, Bugzilla, Mantis, Mindquarry, Redmine, Trac, Bison, CodeSynthesis XSD, CodeSynthesis XSD/e, Flex lexical analyser, Kodos, Open Scene Graph, OpenSCDP, phpCodeGenie, SableCC, SWIG xmlbeansxx, YAKINDU Statechart Tools, Doxygen,

Mkd, Natural Docs, Autoconf, Automake, BuildAMation, CMake, GNU Debugger, Memtest86, Xnee, BOINC, Electric Sheep, XScreenSaver, MyDLP, dvdisaster, Foremost, PhotoRec, TestDisk, Mydiamo, Coyote Linux, Firestarter, IPFilter, ipfw, iptables, M0n0wall, PeerGuardian, PF, pfSense, Rope, Shorewall, SmoothWall, Untangle, Vyatta, Java Astrodynamics Toolkit (GPL), General Mission Analysis Tool (NASA Open Source Agreement), OREKIT (ORbits Extrapolation KIT) (Apache License), Satellite tracking and orbit prediction (GPL), Orbit Reconstruction Simulation and Analysis (GPL), Asteroid Orbit Determination and Propagation (GPL), Libnova (LGPL), Open-Source, Extensible Spacecraft Simulation And Modeling Environment (GPL), Distributed Spacecraft Attitude Control System Simulator (GPL), Solar Sail structure and flight simulator (GPL), SaVi satellite constellation visualizer (BSD License), Rocket Workbench Project (GPL), CpropepShell. Compute propellant performance, BRL-CAD, Blender, Blender CAD, Procad, OpenSCAD, Python CAD, VARKON, OpenCASCADE, FreeCAD, Archimedes, Wikipedia Free CAD Software Listing, Linux.org CAD/CAM Software Listing, NASA Vision Workbench (NOSA license), wikiCalc (GPL), Dia (GPL), DUNE (GPL with runtime exception), Impact (GPL), Code_Aster, SALOME (LGPL), Elmer, Gmsh, OpenFVM, Palabos, Calculix, Package of Additional Octave Libraries, ASCEND modeling environment, OpenDX (IBM), Freshmeat.net Visualization Software Listing, VisIt (BSD), EngLab (GPL), SciLab (CeCILL license), WorldWind (NOSA), Numpy/Scipy, OpenCog, AForge.NET, TREX, ROS, YARP, Blender, flightgear, Chemistry Development Kit, JOELib, CellProfiler, Endrov, FIJI, Ilastik, ImageJ, IMOD ITK, KNIME, VTK, 3DSlicer, Abalone, Ascalaph Designer, GROMACS, LAMMPS, MDynaMix, NAMD, NWChem, Avogadro, BALLView, Jmol, Molekel, MeshLab, PyMOL, QuteMol, RasMol, Ninithi, CP2K, CMU Sphinx, Emacspeak, ESpeak, Bullet, AwayPhysics, Bullet-ANE, ammo.js, Physijs, AmmoNext, Bullet.js, JBullet, ODE, Bounce, nphysics, Velocity Raptor, Box2D, Nape, GoblinPhysics, verlet-js, PhysicsJS, Matter.js, p2.js, Coffee Physics, JPE, APE, Chipmunk2D, glaze, ImpulseEngine, JelloAS3, JelloHx, Jello-Physics, JelloSwift, JigLib, Moby, Newton-Dynamics, OimoPhysics, qu3e, Tokamak, DynaMo, ReactPhysics3D, Chrono Engine, Position Based Dynamics, SPlisHSPlasH etc.

In writing codes that requires open libraries, the modeller may wish to incorporate two or more libraries to suite the desired simulation. For example, a modeller may want a library that can analyze image and at the same extract numerical values that can be plotted. Most image open source application do not plot graph. Hence, if the modeller wishes to plot graph and analyze image at the same time, he needs to incorporate two libraries that can analyze images and plot. This may be very simple; however, it is a difficult task because of the issue of compatibility. Each program has its licenses; hence, the modeller should seek to know the contents of the software licenses to avoid copyright infringement. However, if the modeller, seeks to use the code to solve specific problems (not create application using open source libraries), he may not be infringing on copyright except he commercializes the new program. In the case of creating a new program, the modeller is advised to take keen interest in the statements in the program license. Programs under proprietary licenses cannot

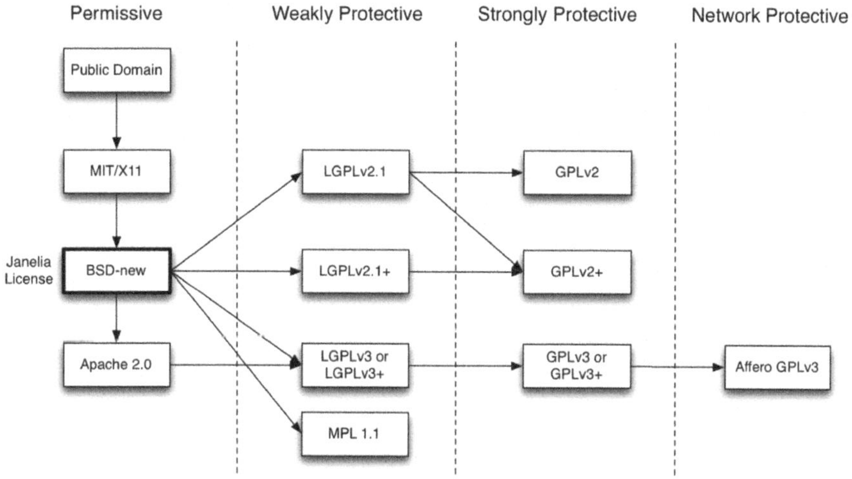

Fig. 3.9 Open source license compatibility chart [6]

be modified or distributed together without violating the licenses. Janelia [6] gave a chart that guides the use of open source licenses (see Fig. 3.9).

References

1. Capterra (2017) Environmental software. https://www.capterra.com/environmental-software/. Accessed on 4 January 2018
2. Patwardhan M (2016) Assessing the impact of usability design features of an mHealth app on clinical protocol compliance using a mixed methods approach. Arizona State University, ProQuest Dissertations Publishing. https://repository.asu.edu/attachments/172769/content/Patwardhan_asu_0010N_16210.pdf
3. Looper J (2016) How to build a great open source community in ten easy steps. https://medium.com/@jenlooper/how-to-build-a-great-open-source-community-in-ten-easy-steps-a3799c1ce52b. Accessed 21 Aug 2018
4. Dodcio (2018) DoD Open Source Software (OSS) FAQ, https://dodcio.defense.gov/Open-Source-Software-FAQ/. Accessed 7 July 2018
5. Haddad I (2015) 7 steps to strengthen your open source compliance. https://s-opensource.org/2015/05/04/7-steps-to-strengthen-your-open-source-compliance/. Accessed 5 July 2018)
6. Janelia F (2018) Open source licenses and their compatibility. http://janelia-flyem.github.io/licenses.html. Accessed 4 Aug 2018

Chapter 4
Modeling Big Data and Further Analysis

As discussed earlier, big data is data indexes that are so huge and complex that conventional data preparing application programming are insufficient to manage them. Big Data can also be defined as the huge volume of data—both organized and unstructured. Big Data is an accumulation of huge data from customary and computerized sources inside and outside an organization that speaks to the validity of an event. Big data is data that contains more prominent assortment touching base in expanding volumes and with ever-higher speed. Big data is trendy expression that is utilized to depict massive volumes of data, both unstructured and organized. Big Data is something that can be utilized to investigate bits of knowledge which can prompt better choices and key research or business moves.

The idea of Big data has been around for quite a long time (Fig. 4.1). Decades before 1950 when the word "big data," was coined, organizations were utilizing fundamental investigation (basically numbers in a spreadsheet that were physically

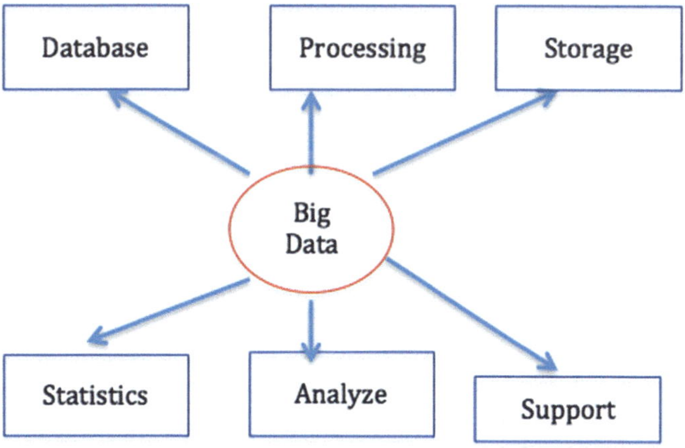

Fig. 4.1 Big data concepts

M. E. Emetere and E. T. Akinlabi, *Introduction to Environmental Data Analysis and Modeling*, Lecture Notes in Networks and Systems 58,
https://doi.org/10.1007/978-3-030-36207-2_4

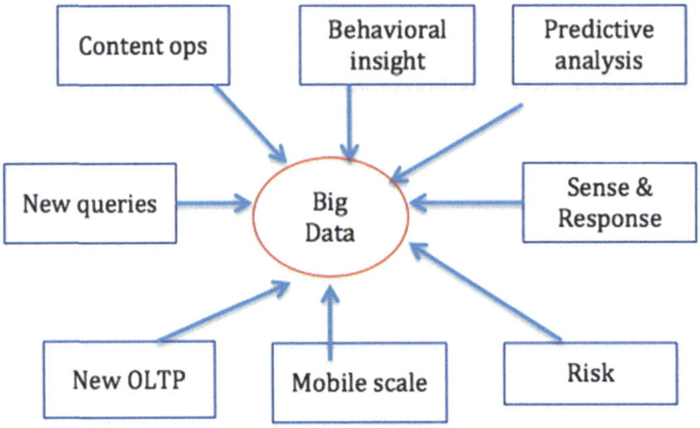

Fig. 4.2 Sub-groups identified under the Big data analytics

inspected) to reveal bits of knowledge and patterns. Thereafter, organization accumulated data that are generated within its system. The "Big Data" idea is ceaselessly developing and being rethought, as it remains the main impetus behind computational advancement, artificial intelligence, data science and the Internet of Things. There are various ideas related with big data which includes volume, variety, velocity, veracity and value. Volume refers to the amount of the dataset, the size of the data set and the potential knowledge accruable from the dataset. Variety is the type and nature of the data that helps a researcher or organization to examine and viably utilize the subsequent understanding. The variety of a big data refers to content, pictures, sound, video etc. Velocity is the rate at which data is produced and handled to address the requests and difficulties that lie in the way of development and improvement. Veracity in big data refers to the variability of the data that influence exact analysis. Value expresses the intrinsic importance Data. However, data is of no utilization until the point when that value is found. Today, big data has turned out to have importance. Hence, the area or fields where big data can be applied shows the value of the data.

Recently, researchers' have reported that big data has a tendency to allude to the utilization of prescient investigation, client conduct examination, or certain other propelled data analytics techniques that concentrate on incentives from data. However, there is little uncertainty on the amounts of data that can be accessed in a computational process. That is the reason big data analytics is quite important as the big data concept. Big data analytics is the utilization of cutting edge systematic methods against extensive, large data sets that incorporate organized, semi-organized and unstructured data, from various sources, and in various sizes from terabytes to zettabytes. The summary of the sub-groups identified under the Big data analytics are presented in Fig. 4.2. The cutting edge systematic methods includes Artificial intelligence (AI), social media, Internet of Things (IoT) etc. Examining Big data

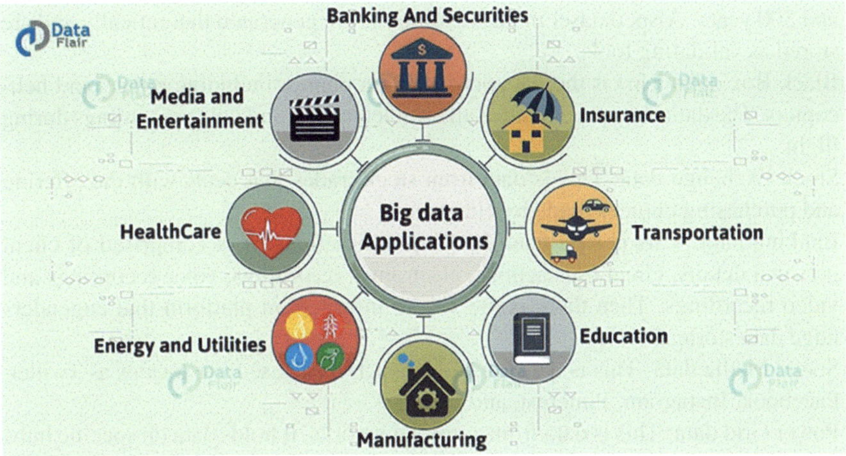

Fig. 4.3 Applications of big data [2]

permits investigators, analysts, and business clients to settle on better and speedier choices utilizing data that was beforehand out of reach or unusable. Big Data Analytics helps researcher to: spots errors in a data; gives room for implementing new methodologies; enhance insights into significant occurrence in the dataset; and reduces cost.

Data Analytics is the art of inspecting crude data with the motivation behind making decisions about that data. Data Analytics includes applying an algorithmic or codes to unravel bits of knowledge. Working with large dataset reveals that the analysis of data sets can discover new relationships to multidisciplinary concept [1]. Hence, "big data" has gained significant relevance in engineering, sciences, business, management, urban planning, media, education, healthcare, manufacturing, international development, information technology, and government. These days, relatively every move we make leaves a computerized trail. We produce data at whatever point we go on the web, when we convey our GPS-prepared cell phones, when we speak with our companions through social media, and when we shop. In this process, the measure of machine-created data is quickly developing as well. Data is produced and shared when our "shrewd" home gadgets speak with each other or with their home servers. In the industry, modern hardware in plants and production lines accumulate and transmit data from sensors. The expression "Big Data" alludes to the accumulation of this data and our capacity to utilize it to further bolster research interest over an extensive variety of zones, including business.

Big data sources represent considerable authority in drawing from various data storehouses and coordinating it into a standard database to make intuitive and adaptable databases for researchers and organizations (Fig. 4.3). The major sources of big data include:

- Research Institutes data: This is the data produced by ground, airplane or satellite measurements that has been observed for a period of time that ranges between 1 and 300 years. Also, dataset in this institutions are generated theoretically and are stored as validating tools.

- Black Box data: This is the data produced via planes, including planes and helicopters. The data incorporates all available electronics and voice recordings during flight.
- Stock Exchange data: This is data from stock trades that deals with the offering and purchasing choices made by clients.
- Banking data: This is data from banking transactions. It is comprised of client personal details, client transactions, electronics recordings, voice recordings and video recordings. Then there is the shared information platform that engenders huge data storage.
- Social Media data: This is data created by such web based life locales as Twitter, Facebook, Instagram, Pinterest, and Google+.
- Power Grid data: This is data from control networks. It holds data on specific hubs and grid systems. It also shares client information, clients' patronage and clients power usage.
- Transport Data: This incorporates conceivable limit, vehicle model, accessibility, and vehicle distance coverage. This data sometimes includes surveillance dataset, voice recordings and client information.
- Search Engine Data: This is one of the greatest wellsprings of big data. Web crawlers have big databases where they get their data. Also, huge electronics resources are linked or shared through this platform.
- Telecommunication data: This data source is adjudged the fastest growing data system because of the voice recordings, client profile, shared information etc.
- School data: This data includes data that has been obtained from primary schools, kindergartens, high schools, colleges and universities. This data includes students, staff and faculty information, research volumes, transactions, electronics data exchange and shared information.
- Embassy data: This area includes visa processing information were lots of information on the traveller is stored for present and future references. Also, there are shared or linked information with sister embassy that results in the possession of huge data.

Some specific big data sources includes Data.gov, Data.gov.uk, NASA.gov, US Census Bureau, Facebook, twitter, Healthdata.gov, Google.com, scopus.com, home.cern etc. The list of big data companies includes Alpine Data Labs, Azure Data Lake, Big Data Partnership, Big Data Scoring, BigPanda, Bright Computing, Cloudera, Compuverde, CVidya, Cybatar, Databricks, Dataiku, DataStax, Domo, Fluentd, Flytxt, Greenplum, GridGain Systems, Groundhog Technologies, Hack/lessen, Hortonworks, HPCC Systems, Imply Corporation, MapR, MarkLogic, Medio, Medopad, Oracle Cloud Platform, Palantir Technologies, Pentaho, Platfora, Quertle, Rocket Fuel Inc., SalesforceIQ, Sense Networks, Shanghai Data Exchange, SocialCops, Sojern, Splunk, Sumo Logic, Talend, Teradata, ThetaRay, TubeMogul, VoloMetrix, Zaloni, Zoomdata etc.

Fig. 4.4 Diagrammatic illustration of the big data analytics processes

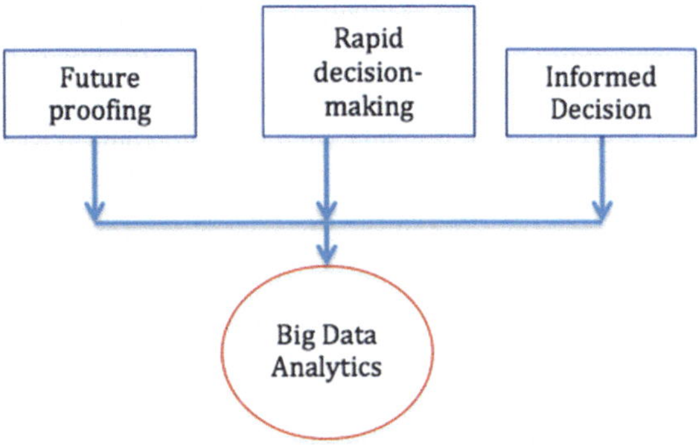

In years to come, data indexes are expected to develop quickly due to progressive accumulation of data generation procedures and Internet of things. Internet of things devices such as cell phones, ethereal (remote detecting), programming logs, cameras, mouthpieces, radio-frequency identification (RFID) and wireless sensor networks are good avenues of increasing the volume of "big data" to hundreds of zettabytes in the nearest future. Big Data deals with the rule that the more you think about anything or any circumstance, the more dependably you can increase new bits of knowledge and influence expectations about the future.

Big Data is all over the place. In business, Big Data results are quite important in organizational choices and vital moves. Hence, some scientists are of the view that conventional data must be incorporated into Big Data since it is a vital bit of the Big Data picture. Lately, there has been a boom in Big Data due to the development of cloud computing, multi-media processing, social media etc. Big organizations have taken the initiates of expanding on data storage mechanism and development of software packages for analyzing big data. Also, big organization are seeking for Big Data solutions to store, process, and break down huge measures of unstructured data.

Big Data solution enables organizations to transform crude data into applicable patterns, forecasts, and projections with exceptional exactness. Big Data Management Solutions enable organizations to accumulate or immerse data from several sources with the aim of improving of client experience. In other words, Big Data analytics is sustained or improved cutting edge Big Data solutions that provides an organization with complete client or customer profiles, empowering the conveyance of more customized client interactions. Big Data may appear to be overwhelming, however with the privilege Big Data solution provides, business and government transaction can be more incisive and exact.

This process is called Big Data analytics. Big Data analytics is the process of examining substantial, huge and shifted data indexes (known as Big data) to reveal concealed examples, obscure connections, market trends, client inclinations and other valuable data that can enable researcher to settle on more-reliable choices. A diagrammatic illustration of the big data analytics processes are presented in Fig. 4.4.

Big Data solutions as touching big data analytics is a lucrative venture. Big data has expanded the request of data information specialists to such an extent that Software AG, Oracle Corporation, IBM, Microsoft, SAP, EMC, HP and Dell have spent more than $15 billion on big data solutions and analytics [3].

For example, big data analytics helps a big organization to:

- identifying the underlying drivers of organizational set-back;
- understanding the capability of data driven choices; executing customer investigation, compliance investigation, fraud investigation, operational investigation;
- gaining new supporters, holding clients, and extending inside current endorser bases are top needs for media transmission specialist organizations;
- gaining new endorsers, holding clients, and growing customer base;
- generating client profile of purchasing propensities;
- improving client commitment through proper documentation of transaction;
- re-evaluating hazard propensities through internal and external data sources;
- archiving client encounter through phone-call, email, WhatsApp and oral engagements;
- fostering time reductions in organization operation;
- understanding the economic situations. For example, organization show signs of improved comprehension of current economic situations. By this, it can stretch out beyond its rivals;
- controlling online notoriety. For example, big data instruments can do sentiments analysis through feedbacks. In this way, organization can get criticism about who is saying what in regards to organization products and services;
- enabling retailers' transaction without much stress via prescient models created from the web based life data, web seek patterns and climate estimates;
- Encouraging organization to make new development openings and completely new classifications of organizations operations that can consolidate on previous profit or gain.

For small businesses or research adventures, big data analytics helps to:

- develop business ideas and choices through dataset that has been gathered online or offline;
- uncover examples, patterns, and affiliations—particularly regarding human conduct and interaction in business. For example, with sales dataset from big company or documented survey, a small business can understand pit-falls to avoid and how to grow their businesses in different geographical setting. In environmental studies, the big data from a satellite source could help the researcher to understand trending phenomenon;
- cut-down on unnecessary expenses. For example, the big dataset from remote source will guide environmental research and save cost of planning research without adequate knowledge on the pitfalls to avoid;
- builds your productivity. Through the experience from existing dataset, research would have enormous opportunity improve upon what is already known;

- enhances researcher or business valuing through resilience to improve upon existing knowledge;
- narrow-down region of interest in research or business;
- foster time reductions in researcher methodology;
- bring-about new product development;
- generate data about the items and administrations, purchasers and providers, buyer inclinations that can be caught and broke down;
- enable small-businesses or researcher to tackle their data and utilize it to recognize new chances. That, thus, prompts more astute business or research moves, more productive tasks, higher benefits and more joyful clients or results.

Big Data gives us uncommon bits of knowledge and openings, yet it likewise brings up concerns and issues that must be tended to. This includes: Data protection—the Big Data we currently create contains a considerable measure of data about our own lives, a lot of which we have a privilege to keep private. Progressively, we are requested to strike a harmony between the measure of individual data we unveil, and the accommodation that Big Data solutions offer; Data security—to what extent can a data be protected when it receives the approval of the organization, researcher or personality.

Aside, the above mentioned problems, Big data challenges incorporate catching data, data stockpiling, data analytics, look, sharing, exchange, representation, questioning, refreshing, data security and data source. One of the challenges of big data is to process and interpret it, such that, the researcher identifies challenges that need further work or answers. So, big data solution is all about accessibility to use or coordinate what a researcher or organization had at the top of the priority list for estimating the level of successes recorded. The common challenge researcher face is to overpowered by every one of the data that is out there due to inefficient analytics tools or software. Hence, the more effectively an organization or utilizes its data, the more potential it needs to develop. The organization can take data from any source and dissect it to discover answers to bugging challenges.

Data must be handled with cutting edge instruments to uncover the significance of the data. For instance, to deal with air pollution research, one must consider both pollution source and meteorological issues with different segments. Hence, importance must be placed on the cutting edge instrument or software that helps the individual to accomplish his/her task. Data Analysis have found importance in healthcare, travel, gaming, energy management etc. Unstructured and semi-organized data types ordinarily do not fit well in traditional data warehouses that are based on relational databases situated to organized data sets. Moreover, data storage (looking at the volume of data set that is generated in recent time) may be unable to deal with the handling requests presented by sets of Big Data. Hence, the tools used to analyze big data includes: NoSQL databases, Hadoop, YARN, MapReduce, Spark, HBase, Hive, Kafka, Pig, Earthenware, Avro, Oozie, Tempest, Cloudera, Gridgrain, SpaceCurve, SpaceCurve, Domo, Teradata Database, Hitachi Vantara, Statistica, Panoply, IBM Watson Analytics, SAS Visual Analytics, Sisense, Talend, MongoDB, Vertica

Analytics Platform, SAP Vora, Oracle Big Data suite, Apache Cassandra, Plotly, Wolfram Alpha, Tibco Spotfire, AnswerRocket, Tableau, Zookeeper etc.

In some cases, Hadoop clusters and NoSQL frameworks are being utilized basically to prepare data before it gets stacked into a data distribution center or scientific database for investigation, more often than not in an abridged shape that is more helpful for social structures. Once the data is prepared, it may be investigated with the product usually utilized in advanced analytics processes. That incorporates devices for data mining, which filter through data indexes looking for designs and relationships; predictive investigation (which fabricate models for anticipating client conduct and other future improvements); machine learning which use set of codes to break down vast data sets. Big data can be analyzed using programming language, for example, R, Python, Scala, C, C++, and SQL. Earthenware enables venture applications to store and oversee Big Data in server memory, significantly speeding execution. Apache Avro is an data serialization framework in view of JSON-characterized patterns. Zookeeper and Oozie uses the planning of Hadoop framework. Tempest is an ongoing disseminated calculation framework. It works indistinguishable route from Hadoop to analyze data. Cloudera offers continuous, intelligent diagnostic questions of the data put away in HBase or HDFS. GridGain is used with Hadoop DFS and it offers a substitute to Hadoop's MapReduce. SpaceCurve is used for multidimensional geodata. Domo is used to give organizations an approach to imagine their data from various and different data sources. Teradata Database has new Big Data abilities like the Teradata Unified Data Architecture, empowering organizations to access and process scientific inquiries over various frameworks, including bi-bearing data import and fare from Hadoop. Hitachi Vantara is used for systems stockpiling and server farm framework business. Statistica is used to perform data mining on organized and unstructured data, addresses IoT data, can convey investigation on gadgets and entryways anyplace on the planet. Panoply uses AI to dispense with the advancement and coding required for changing, incorporating, and overseeing data.

4.1 Description of Data Sources

The beauty of environmental research is that data can be found everywhere and lot of the found data are actually free (As seen in Appendix). The moves by organizations and governments to put a lot of data in public domain have engendered huge volumes of data open to everybody. In this section, shall highlight some of the environmental data sources and the web link to those datasets. Why is the source link important in environmental studies?

- It enables institutions to share information.
- It enables researcher to get access to these data to corroborate, validate or establish models.
- It leads to the generation of more dataset in form of predictive models or optimization models.

- It reduces the cost of planning experiment.
- It serves as a log-book for knowing what type of experiments have been done and what geographical area it can be found.
- It enables comparative study of same event in different locations.
- It is the simplest way to contribute to research in developing countries.

The available data sources are listed below.

i. **DATA.GOV**—Environmental Data Sources (https://catalog.data.gov/dataset/environmental-data-sources). This database incorporates measuring stations, climatic server farms, and storet destinations. The sources of its information are sourced from: KGSWL—Water level wells in the KGS and USGS databases; National Climatic Data Center (NCDC)—Weather information gathering locations; EPAKS—All STORET (EPA database) that accounts for STORET database in 25 locations. STORET is a user-maintained database; USGSSW—USGS Surface Water measuring/gauging stations; NOAA—where the U.S. Hourly Precipitation Data can be found.
ii. **UN Environment**—has different sites where environmental data can be found. The State of Food and Agriculture can be found in the link—http://www.fao.org/faostat/en/. This site has the most recent dataset of seventy-four countries. The second site borders on the Advancement on sanitation and drinking—drawn from MICS and DHS studies in 25 sub-Saharan African nations and other parts of the world (http://mics.unicef.org/surveys). The third environment site in the ambient air pollution dataset for the continents of the world. It can be found in the web link—http://www.who.int/phe/health_topics/outdoorair/databases/AAP_BoD_results_March2014.pdf?ua=1.
iii. **Centre for Environmental Data Analysis (CEDA)** is a United Kingdom association that serves the environmental science network by arrangement of data center, information examination, information access and research venture interest. CEDA obligation incorporates three datacentres mutually supported by the Natural Environment Research Council (NERC) and the Science and Technology Facilities Council (STFC). CEDA is in charge of giving access to investigate data. The link to the data is given as http://www.ceda.ac.uk/.
iv. **Chinese Ecosystem Research Network (CERN)** is an ecosystem research network with field stations throughout China covering the fields of agriculture, forest, grassland, lake and marine ecosystems. The link to the data is given as http://www.cern.ac.cn/0index/index.asp. The objectives of the CERN include

    - To lead long haul observing projects for biological systems in horticulture, ranger service, prairie, lakes and sounds. Also it observes ecological components, similar to water, soil, and air.
    - To direct inside and out examinations of the structure, function, and elements of the environments, and approaches and strategies for supportable use of these frameworks.

- To give ideal administrative models to nearby communities for the maintainable usage of the nearby regular resources, and change of the nearby living condition.
- To give logical bases to leaders at different managerial levels with respect to the use of regular assets and condition.
- To add to the arrangement of worldwide biological and environmental issues through dynamic cooperation in international biological research ventures.

v. **UK Eutrophying and Acidifying atmospheric Pollutants (UKEAP)** network measures air pollutants at rural sites across the UK. It comprises of five components i.e. Precipitation network (to acquire inorganic anions & cations concentrations in precipitation dataset for 39 locations); National Ammonia Monitoring Network (is used to obtain monthly gas phase ammonia concentrations dataset for 85 locations); Acid Gas and Aerosol Network (to acquire monthly gas phase SO_2, HNO_3, HCl dataset for major particulate phase inorganic anions and cations in 30 locations); NO2Net (to acquire NO_2 concentrations dataset in 24 locations). UKEAP principally seeks to screen eutrophying and acidifying species in air and precipitation in country territories of the United Kingdom with adequate spatial and fleeting goals to permit: the assessment of approach measures to lessen fixation and statement; dangers to biological systems and complications of basic burdens to be evaluated; evaluations of auxiliary airborne for PMx. The link to the data is given as http://www.pollutantdeposition.ceh.ac.uk/ukeap.

vi. **UK Environmental Observation Framework** is a United Kingdom association that works to improve coordination of the observational evidence needed to understand and manage the changing natural environment. Some of the specific operation of the organization includes: technological landscape, satellite imagery observation, thermal Imaging in ecology, tracking and tagging environmental changes, information sharing with National Biodiversity Network, eDNA and Plant Pathogen Metabar coding, and capturing and ingestion methods for geoscientific data. The link to the data is given as http://inspire.ec.europa.eu/Themes/Data-Specifications/2892.

vii. **Climatic Research Unit (CRU)** is hosted in University of East Anglia UK made accessible gridded datasets of surface temperature information over land zones and midpoints for the Northern and Southern Hemispheres and the Globe. These datasets (the most recent being CRUTEM4, http://www.cru.uea.ac.uk/cru/information/temperature/) have been created from information gained from climate stations in the globe. All these climate stations are controlled by National Meteorological Services (NMSs) who trade this information over the CLIMAT arrange that is a piece of the World Meteorological Organization's (WMO) Global Telecommunications System (GTS). CRU also seeks to discover extra, and homogenized, information from an extensive variety of sources. This is evident in the advancement of gridded datasets and the station information document throughout the years. The techniques adapted by CRU to build-up the gridded datasets have been portrayed in various publications found on the web link http://www.cru.uea.ac.uk/cru/information/temperature/and the connected FAQs.

viii. **Data Distribution Centre (DDC)** of the Intergovernmental Panel on Climate Change (IPCC) is overseen by the IPCC Task Group on Data and Scenario Support for Impact and Climate Analysis (TGICA) and jointly managed by the British Atmospheric Data Centre (BADC) in the United Kingdom, the ICSU World Data Center Climate (WDCC) in Germany, and the Center for International Earth Science Information Network (CIESIN) at Columbia University, New York, USA. The activities of the DDC includes

- Observations of the climate
- Socio-Economic Baseline Dataset
- IPCC AR5 Observed Climate Change Impacts
- IPCC AR4 Observed Climate Change Impacts
- Carbon Dioxide: observed atmospheric concentrations.

The link to the data is given as

ix. **National Center for Atmospheric Research (NCAR)** managed by the non-profit University Corporation for Atmospheric Research (UCAR) and funded by the National Science Foundation (NSF) was intended to furnish the climatic research network with the mutual assets important to take a shot at the most imperative logical issues of the day. The job specification of NCAR include:

- specialized instruments to measure atmospheric processes,
- Research aircraft,
- High-performance computing and cyber infrastructure, including supercomputers,
- Mauna Loa Solar Observatory,
- Cooperative field campaigns,
- Atmospheric models of weather, chemical, solar, and climate processes (Community Climate System Model (CCSM); Weather Research and Forecasting model (WRF); Whole Atmosphere Community Climate Model (WACCM)),
- Technology transfer to support societal needs,
- Data sets, data services, and other resources,
- NCAR Command Language (NCL), a programming language designed for use with climate and model data. The data can be obtained from https://data.ucar.edu/.

x. **NOAA Earth System Research Laboratory** is a lab in National Oceanic and Atmospheric Administration's (NOAA) Office of Oceanic and Atmospheric Research (OAR). It is one of seven NOAA Research Laboratories (RLs) and is arranged in Boulder, Colorado. ESRL unites six separate NOAA labs under one relationship with four Divisions—Global Monitoring, Physical Sciences, Chemical Sciences, and Global Systems. The past labs falling into

ESRL are the Aeronomy Laboratory, the Climate Diagnostics Center, the Climate Monitoring and Diagnostics Laboratory, the Environmental Technology Laboratory, the Forecast Systems Laboratory, and the Surface Radiation Research Branch of the Air Resources Laboratory. NOAA Earth System Research Laboratory is connected to several dataset listed below.

- **CMAP Precipitation**
  Monthly and pentad global gridded precipitation means. It includes a standard and enhanced version (with NCEP Reanalysis) from 1979 to near the present. It can be linked through the web vis https://www.esrl.noaa.gov/psd/data/gridded/data.cmap.html. The monthly information index comprises of two documents containing monthly averaged precipitation rate estimates. Qualities are acquired from 5 sorts of satellite evaluations (GPI, OPI, SSM/I dispersing, SSM/I discharge and MSU) and gauge dataset.
- **COBE-SST**
  Monthly 1 × 1 SST dataset from 1891 to present from the JMA. Datasets uses 3D Var to fill gaps. The dataset can be obtained from the link https://www.esrl.noaa.gov/psd/data/gridded/data.cobe.html. **The COBE-SST dataset** is utilized as contribution for the JMA Climate Data Assimilation System (JCDAS) and the Japanese 25-year Re-investigations (JRA-25 and JRA-55). It is likewise utilized as contribution for the gathering expectation frameworks utilized in regular estimates.
- **COBE-SST2 Sea Surface Temperature**
  Monthly 1 × 1 SST dataset from 1850 to present from the JMA. The dataset is found on this web link https://www.esrl.noaa.gov/psd/data/gridded/data.cobe2.html.
- **CPC.25x.25 Daily US Unified Precipitation**
  High resolution gridded precipitation over the US (from station data) for 1948–2006 extending to the present for a real-time version. The dataset can be obtained from the link. https://www.esrl.noaa.gov/psd/data/gridded/data.unified.daily.conus.html. The essential objective of the project is to make suite of standard precipitation items with reliable amount and enhanced quality by joining all data sources accessible at CPC and by exploiting the optimal interpolation (OI) target investigative procedure.
- **CPC Hourly Precipitation**
  NCEP's hourly gridded US station precipitation from 1948. The dataset can be obtained from the web link https://www.esrl.noaa.gov/psd/data/gridded/data.cpc_hour.html.
- **CPC Soil Moisture**
  Monthly Gridded CPC Soil Moisture from a model from 1948 to present. The dataset can be obtained from the web link https://www.esrl.noaa.gov/psd/data/gridded/data.cpcsoil.html. The monthly data collection comprises of the document containing monthly averaged soil moisture water height equivalents. Note that the information is calculated and not estimated specifically. The dataset is presently V2. There are a few contrasts with

the past adaptation, especially over Africa. The V2 form likewise has the land mask connected to the data values.

- **CRU Air Temperature and Combined Air Temperature/Marine Anomalies V4**
  Global gridded (5° × 5°) monthly anomalies of observed air temperature and combined observed air and marine temperature (HADCRU4) from the mid 1800s to near present. Current version is V4.2.0 which was put out 5/9/2013. The dataset can be obtained from the web link https://www.esrl.noaa.gov/psd/data/gridded/data.crutem4.html.
- **Dai Palmer Drought Severity Index**
  Monthly gridded global PDSI (Palmer Drought Severity Index) values from 1850 to 2010 from NCAR. The dataset can be obtained from the web link https://www.esrl.noaa.gov/psd/data/gridded/data.pdsi.html. **The Palmer Drought Severity Index (PDSI)** is devised by Palmer [4] to represent the severity of dry and wet spells over the U.S. based on monthly temperature and precipitation data as well as the soil-water holding capacity at that location.
- **GFS Model Output**
  A dataset of historical weather forecasts generated with a fixed numerical model. Daily forecasts are comprised of a 15 day, 15-member ensemble and include surface and pressure level variables. The dataset can be obtained from the web link https://www.esrl.noaa.gov/psd/forecasts/reforecast2/.
- **GHCN Version 3 Land Temperature Dataset**
  GHCN V3 contains gridded temperature anomalies calculated from the GHCN version 3 monthly temperature data set. The dataset can be obtained from the web link https://www.esrl.noaa.gov/psd/data/gridded/data.ghcngridded.html. The temperature data set contains gridded mean temperature anomalies from the Global Historical Climatology Network-Monthly (GHCN-M) version 3.2.1 temperature data set.
- **GHCN version 2 Land Precipitation Dataset**
  GHCN V2 contains gridded precipitation anomalies calculated from the GHCN version 2 monthly precipitation data set. 2,064 homogeneity adjusted precipitation stations were combined with a data set containing 20,590 raw precipitation stations throughout the world onto a 72 × 36 (5 deg × 5 deg) global grid. The dataset can be obtained from the web link https://www.esrl.noaa.gov/psd/data/gridded/data.ghcncams.html.
- **NOAA GHCN_CAMS Land Temperature Analysis**
  GHCN CAMS is a high resolution (0.5 × 0.5) analyzed global land surface temperatures from 1948 to near present. The dataset can be obtained from the web link https://www.esrl.noaa.gov/psd/data/gridded/data.gistemp.html.
- **GISS surface temperature analysis**
  Goddard Institute for Space Studies (GISS/NASA) surface temperature analysis for the globe; globally gridded at a 2 × 2 resolution. The dataset can

be obtained from the web link https://www.esrl.noaa.gov/psd/data/gridded/data.gistemp.html.

- **NCEP GODAS ocean analysis**
  High Resolution Multi-level ocean analysis from NCEP. The dataset can be obtained from the web link https://www.esrl.noaa.gov/psd/data/gridded/data.godas.html. GODAS is a real-time ocean analysis and a reanalysis. It is used for monitoring, retrospective analysis as well as for providing oceanic initial conditions for the CFS.
- **Global Precipitation Climatology Centre (GPCC)**
  GPCC Global Precipitation Climatology Centre monthly precipitation dataset from 1901–present is calculated from global station data. The dataset can be obtained from the web link https://www.esrl.noaa.gov/psd/data/gridded/data.gpcc.html.
- **GPCP V2.3 Precipitation**
  Global Precipitation Climatology Project monthly precipitation dataset from 1979–present combine observations and satellite precipitation data into $2.5^\circ \times 2.5^\circ$ global grids. The dataset can be obtained from the web link https://www.esrl.noaa.gov/psd/data/gridded/data.gpcp.html.
- **ICOADS**
  Global surface marine data from 1800 to near the present summarized in monthly gridded formats ($2^\circ \times 2^\circ$ boxes, or $1^\circ \times 1^\circ$ boxes from 1960 forward), and offering a variety of statistics. The dataset can be obtained from the web link http://icoads.noaa.gov/data.icoads.html.
- **Interpolated OLR**
  Gridded daily and monthly OLR data from NCAR with temporal interpolation. See the related not interpolated OLR dataset. The dataset can be obtained from the web https://www.esrl.noaa.gov/psd/data/gridded/data.interp_OLR.html.
- **Kaplan SST**
  Gridded global SST anomalies from 1856–present derived from UK Met Office SST data which has had sophisticated statistical techniques applied to it to fill in gaps. The dataset can be obtained from the web https://www.esrl.noaa.gov/psd/data/gridded/data.kaplan_sst.html. This data set is produced by taking MOHSST5 version of the GOSTA data set from the U.K. MET office as the input SST data set to various processing steps.
- **Livneh daily CONUS near-surface gridded meteorological and derived hydrometeorological data**.
  This CONUS daily dataset from 1915 to 2011 is 1/16 resolution. The dataset variables have been generated using the Variable Infiltration Capacity VIC hydrologic model v.4.1.2.c which was driven with the companion meteorological data. The dataset can be obtained from the web https://www.esrl.noaa.gov/psd/data/gridded/data.ncep.html.
- **NCEP Operational Analysis**
  NCEP's twice-daily global analysis at 2.5° resolution on pressure levels which is a product of their operational forecast system. The dataset can be

obtained from the web https://www.esrl.noaa.gov/psd/data/gridded/data.ncep.marine.html. PSD has retired this dataset as of Sep 7th 2016.

- **NCEP Marine**
  Marine data gathered by NCEP summarized into ICOADS-compatible (2° × 2° box) statistics. Production of this dataset at PSD was phased out effective in Mar. 2011 (thus data cover 1991–Feb. 2011). The dataset can be obtained from the web https://www.esrl.noaa.gov/psd/data/gridded/data.ncep.reanalysis.html.
- **NCEP/NCAR Reanalysis**
  A state-of-the-art analysis/forecast system is used to perform data assimilation using data from 1948 to the present. The dataset can be obtained from the web https://www.esrl.noaa.gov/psd/data/gridded/data.ncep.reanalysis.derived.html.
- **NCEP/NCAR Reanalysis Products Derived at PSD**
  A subset of the NCEP/NCAR Reanalysis has been processed to create monthly means of the original data, and in some cases, derived variables or other statistics. The dataset can be obtained from the web https://www.esrl.noaa.gov/psd/data/gridded/data.ncep.reanalysis2.html.
- **NCEP/DOE Reanalysis II**
  A state-of-the-art analysis/forecast system is used to perform data assimilation using data from 1979 through 2003. A large subset of this data is available from PSD in its original 4 times daily format and as daily averages. The dataset can be obtained from the web https://www.esrl.noaa.gov/psd/data/gridded/data.ncep.reanalysis2.html. The NCEP-DOE Reanalysis 2 project is utilizing the best investigation/figure framework to perform data assimilation utilizing past information from 1979 through the earlier year. An extensive subset of this information is accessible from PSD 4 times every day and as daily averages.
- **NOAA Extended Reconstructed SST V3b**
  A global monthly SST analysis from 1854 to the present derived from ICOADS data with missing data filled in by statistical methods. The dataset can be obtained from the web https://www.esrl.noaa.gov/psd/data/gridded/data.noaa.ersst.v3.html.
- **NOAA Extended Reconstructed SST V4**
  A global monthly SST analysis from 1854 to the present derived from ICOADS data with missing data filled in by statistical methods. The dataset can be obtained from the web https://www.esrl.noaa.gov/psd/data/gridded/data.noaa.ersst.v4.html.
- **NOAA Extended Reconstructed SST V5**
  A global monthly SST analysis from 1854 to the present derived from ICOADS data with missing data filled in by statistical methods. The dataset can be obtained from the web https://www.esrl.noaa.gov/psd/data/gridded/data.noaa.ersst.v5.html. The Extended Reconstructed Sea Surface Temperature (ERSST) dataset is a global monthly sea surface temperature dataset

derived from the International Comprehensive Ocean–Atmosphere Dataset (ICOADS).

- **NOAA's Outgoing Longwave Radiation–Daily Climate Data** Record (OLR–Daily CDR): PSD Interpolated Version
  Gridded daily 1 × 1 OLR CDR data from NCDC interpolated to −90 to 90. The dataset can be obtained from the web https://www.esrl.noaa.gov/psd/data/gridded/data.olrcdr.interp.html.
- **NOAA High-resolution Blended Analysis of Daily SST and Ice.**
  Data is from Sep 1981 and is on a 1/4 deg global grid. The dataset can be obtained from the web https://www.esrl.noaa.gov/psd/data/gridded/data.noaa.oisst.v2.highres.html.
- **NOAA Global Surface Temperature (NOAAGlobalTemp)**
  Monthly gridded anomalies of air temperature from the GHCN-M and the NOAA ERSST V4 dataset. The dataset can be obtained from the web https://www.esrl.noaa.gov/psd/data/gridded/data.noaaglobaltemp.html. The NOAA Global Surface Temperature Dataset (NOAAGlobalTemp) is a merged land–ocean surface temperature analysis (formerly known as MLOST) (link is external). It is a spatially gridded ($5^\circ \times 5^\circ$) global surface temperature dataset, with monthly resolution from January 1880 to present.
- **NOAA Optimum Interpolation (OI) SST V2**
  Produced weekly on a 1° grid. The analysis uses in situ and satellite SST's plus SST's simulated by sea-ice cover. Before the analysis is computed, the satellite data is adjusted for biases. Monthly and monthly long-term means are also available. The dataset can be obtained from the web https://www.esrl.noaa.gov/psd/data/gridded/data.noaa.oisst.v2.html.
- **NOAA Highly Reflective Clouds**
  Subjectively identified areas of large-scale organized convection over the global tropics have been gridded daily for a 15-year period from 1971 to 1988. The dataset can be obtained from the web https://www.esrl.noaa.gov/psd/data/gridded/data.noaa.hrc.html.
- **NODC (Levitus) World Ocean Atlas 1994**
  Annual, seasonal and monthly long term means for temperature and salinity at multiple depth. Several statistics such as number of observations are included in separate files. The dataset can be obtained from the web https://www.esrl.noaa.gov/psd/data/gridded/data.nodc.woa94.html.
- **NODC (Levitus) World Ocean Atlas 1998**
  The atlas contains annual, seasonal, and monthly long term means for temperature and salinity at multiple depths. Several statistics, including nobs are included in separate files, along with two "mask" files. The dataset can be obtained from the web https://www.esrl.noaa.gov/psd/data/gridded/data.nodc.woa98.html.
- **NOAA's Precipitation Reconstruction (PREC)**
  The dataset can be obtained from the web. https://www.esrl.noaa.gov/psd/data/gridded/data.prec.html. The PREC data set is an analysis of monthly

precipitation constructed on a 2.5(^{o}) lat/long grid over the global for the period from 1948 to the present. The land portion of this global analysis is defined by optimum interpolation of gauge observations at over 17,000 stations collected in the NOAA/NCDC GHCN Version 2 and the NOAA/CPC CAMS data sets.

- **NOAA's Precipitation Reconstruction over Land (PREC/L)**
  It can be linked through https://www.esrl.noaa.gov/psd/data/gridded/data.precl.html. The monthly data set consists of three files containing monthly averaged precipitation totals. Precipitation is available at 3 spatial resolutions.
- **North American Regional Reanalysis (NARR)**
  NCEP's high resolution combined model and assimilated dataset. It covers 1979 to near present and is provided 8-times daily, daily and monthly on a Northern Hemisphere Lambert Conformal Conic grid for all variables. The dataset can be obtained from the web https://www.esrl.noaa.gov/psd/data/gridded/data.narr.html.
- **Northern Hemisphere EASE-Grid Snow Cover and Sea Ice Extent**
  The NSIDC weekly snow cover and ice dataset has been interpolated to a $1^\circ \times 1^\circ$ Northern Hemisphere grid and a monthly time-scale covering 1971–1995. The dataset can be obtained from the web https://www.esrl.noaa.gov/psd/data/gridded/data.snowcover.html.
- **NOAA-CIRES 20th Century Reanalysis (V2)**
  20th Century Reanalysis contains objectively-analyzed 4-dimensional weather maps and their uncertainty from the late 19th century to 21st century. The dataset can be obtained from the web https://www.esrl.noaa.gov/psd/data/gridded/data.20thC_ReanV2.html.
- **NOAA-CIRES 20th Century Reanalysis (V2c)**
  NOAA-CIRES 20th Century Reanalysis V2c contains objectively-analyzed 4-dimensional weather maps and their uncertainty from the mid 19th century to 21st century. The dataset can be obtained from the web https://www.esrl.noaa.gov/psd/data/gridded/data.20thC_ReanV2c.html.
- **University of Delaware Precipitation and Air Temperature**
  Monthly global gridded high resolution station (land) data for air temperature and precipitation from 1900 to 2014. The dataset can be obtained from the web https://www.esrl.noaa.gov/psd/data/gridded/data.UDel_AirT_Precip.html.
- **Uninterpolated OLR**
  Gridded OLR data from NCEP without interpolation. See the Interpolated OLR dataset for an interpolated version. The dataset can be obtained from the web https://www.esrl.noaa.gov/psd/data/gridded/data.uninterp_OLR.html.
- **CPC Global Precipitation**
  Daily Gridded CPC Precipitation for 1979 to present. The dataset can be obtained from the web https://www.esrl.noaa.gov/psd/data/gridded/data.cpc.globalprecip.html. His data set is part of products suite from the CPC

Unified Precipitation Project that are underway at NOAA Climate Prediction Center (CPC). The primary goal of the project is to create a suite of unified precipitation products with consistent quantity and improved quality by combining all information sources available at CPC and by taking advantage of the optimal interpolation (OI) objective analysis technique.

- **CPC Global Temperature**
  Daily Gridded CPC Temperature from the GTS for 1979 to present. The dataset can be obtained from the web https://www.esrl.noaa.gov/psd/data/gridded/data.cpc.globaltemp.html.
- **NCEP/NCAR Reanalysis I (1948–present)**
  This reanalysis was the first of its kind for NOAA. NCEP used a comparable climate model that were instated with a wide variety of atmosphere recognitions: ships, planes, RAOBS, station data, satellite observations and some more. By using a comparable model, scientists can take a gander at environment/atmosphere estimations and dynamic strategies without the trouble that model changes can cause. The dataset is kept current using close constant observations. The dataset can be obtained from the web https://www.esrl.noaa.gov/psd/data/reanalysis/reanalysis.shtml.
- **NCEP/DOE Reanalysis II (1979-display)**
  NCEP conveyed a second type of their first reanalysis starting from the most punctual beginning stage of the genuine satellite time. More observations were incorporated, absorption botches were amended and a predominant version of the model was used. The dataset can be obtained from the web https://www.esrl.noaa.gov/psd/data/gridded/data.ncep.reanalysis2.html.
- **NARR (1979-appear)**
  The NARR reanalysis was done to convey high objectives yield over the North American territory. Observational information sources resembled NCEP I with the extension of assimilated precipitation. The NARR exhibit region was settled in an around the world, cut down objectives appear. Yields resemble the NCEP I and II models anyway with more snow, ice and precipitation related elements. The dataset can be obtained from the web https://www.esrl.noaa.gov/psd/data/narr/.

xi. **Climatic Data OnLine (CDO)** Climate Data Online (CDO) provides free access to NCDC's archive of global historical weather and climate data in addition to station history information. These data include quality controlled daily, monthly, seasonal, and yearly measurements of temperature, precipitation, wind, and degree days. The data can be found on the web link https://www7.ncdc.noaa.gov/CDO/cdo.

xii. **PRISM Climate Data** offers climate observations from a wide range of monitoring networks, applies sophisticated quality control measures, and develops spatial climate datasets to reveal short- and long-term climate patterns. The resulting datasets incorporate a variety of modeling techniques and are available at multiple spatial/temporal resolutions, covering the period from 1895 to the present.

xiii. **WorldClim—Global Climate Data** a set of global climate layers (gridded climate data) with a spatial resolution of about 1 km^2. These data can be used for mapping and spatial modeling. The new Version 2.0 is also available. The data can be found on the web link http://www.worldclim.org/.

xiv. **SSURGO/STATSGO2**—Fares from the Soil Data Mart are conveyed in what is alluded to as Soil Survey Geographic Database (SSURGO) arrange. The accompanying archives, outlines, and reports portray the SSURGO standard and STATSGO2, and additionally the devices and strategies that are important to viably utilize this information. The Digital General Soil Map of the United States or STATSGO2 is an expansive based stock of soils and non-soil zones that happen in a repeatable example on the scene and that can be cartographically appeared at the scale mapped of 1:250,000 in the mainland U.S., Hawaii, Puerto Rico, and the Virgin Islands and 1:1,000,000 in Alaska. The dataset can be found in the web link https://sdmdataaccess.nrcs.usda.gov/.

xv. **Global High-Resolution Soil-Water Balance** dataset gives hydrological raster information (ESRI Grid design) depicting real evapotranspiration and soil water deficiency with the goals of 30 circular segment seconds (~920 m at the equator). This Soil-Water Balance uses the WorldClim and Global-PET database as essential information. The technique used to infer these datasets is depicted in the dataset documentation. The outcomes feature particularly the climatic effect on hydrological measurements controlling vegetation reasonableness. This dataset depends on displaying and examinations by Antonio Trabucco (Forest Ecology and Management Research Group, K.U. Leuven), with the help of the International Water Management Institute (IWMI) and the International Center for Integrated Mountain Development (ICIMOD), and are given online by the CGIAR-CSI Consortium for Spatial Information with the help of the International Center for Tropical Agriculture (CIAT). The dataset can be found in the web link https://cgiarcsi.community/data/global-high-resolution-soil-water-balance/.

xvi. **The Global Potential Evapo-Transpiration (Global-PET) and Global Aridity Index (Global-Aridity)** dataset give high-goals worldwide raster atmosphere information identified with evapotranspiration procedures and precipitation shortfall for potential vegetative development. These datasets depend on displaying and investigations by Antonio Trabucco (CSI-Alumni/at present at the Euro-Mediterranean Center on Climate Change, Italy), with the help of the International Water Management Institute (IWMI) and the International Center for Integrated Mountain Development (ICIMOD), and are furnished online by the CGIAR-CSI with the help of the International Center for Tropical Agriculture (CIAT). The Global-PET and Global-Aridity datasets are accommodated non-business use in standard ARC/INFO Grid design, at 30 circular segment seconds (~1 km at the equator), to help contemplates adding to supportable advancement, biodiversity and natural preservation, neediness lightening, and adaption to environmental change all inclusive, and specifically in creating nations. The dataset can be found in the web link https://cgiarcsi.community/data/global-aridity-and-pet-database/.

xvii. **EOS-Earth Data** provides free, customized Earth Science data. All data are freely available. It has over 1,000,000 data products that to our users world-wide since 1999. It has three main sections i.e. satellite data, regional and global projects and geographic and demographic data. Satellite Data includes: Educational Image Subsets; Global Vegetation Indices; and MODIS Aggregate Products. The Regional and Global Projects include: Carbon Cycle Model Linkage-CCMLP; Climate Changes in the 21st Century; NCAR Science and Research Data for IPCC AR4; US Hydrologic Data (GHAAS); US National Assessment; VEMAP2 Transient Dynamics Data; and VEMAP2 Transient Dynamics Data for Alaska. The Geographic and Demographic Data include: China County Data; Global Demographic Data; Global Land use Modeling; Global Model Reference Data; India Rice Cropping Maps; LBA HydroNet Data; LBA/South American Data; New England Methane; New Hampshire Land cover; Tapajos Tree Census; and US County Data. The dataset can be found in the link http://eos-earthdata.sr.unh.edu/data/.

xviii. The **Forest Inventory and Analysis (FIA)** Program of the U.S. Forest Service gives the data expected to evaluate America's woodlands. As the Nation's constant woodland statistics, our program extends how forests are probably going to show up 10 to quite a while from now. This empowers us to assess whether current forests administration hones are maintainable over the long haul and to evaluate whether current approaches will enable the cutting edge to make the most of America's forests as we do today. Its dataset can be found on the web link https://apps.fs.usda.gov/fia/datamart/recent_load_history.html.

xix. **NHDPlus** is a geo-spatial, hydrologic structure dataset worked by the US EPA Office of Water, helped by the US Geological Survey. Since its first discharge in 2006, NHDPlus has been made accessible to the more extensive water assets network and has been utilized for some differing applications by Federal, state, and nearby governments, and in addition non-benefit organizations, privately owned businesses and instructive foundations. The NHDPlus client network has created instruments and ancillary information to help their NHDPlus applications. A portion of these applications are recorded here. NHDPlus is a coordinated suite of utilization prepared geospatial informational collections that consolidate huge numbers of the best highlights of the National Hydrography Dataset (NHD), the National Elevation Dataset (NED), and the Watershed Boundary Dataset (WBD). NHDPlus Version 1 (NHDPlusV1) was discharged in 2006. NHDPlus Version 2 (NHDPlusV2) started its open introduction in June 2012. Both NHDPlusV1 and V2 consolidate the NHD 1:100,000-scale stream arrange and the 1 curve second NED (roughly 30 m ground dispersing). Its dataset can be found on the web link http://www.horizon-systems.com/NHDPlus/NHDPlusV2_data.php

xx. **GlobCover** is an ESA initiative which started in 2005 in association with JRC, EEA, FAO, UNEP, GOFC-GOLD and IGBP. The point of the project was to build up an administration equipped for conveying worldwide composites and land cover maps utilizing as information perceptions from the 300 m MERIS sensor on board the ENVISAT satellite mission. ESA makes accessible the land cover maps, which cover 2 periods: December 2004–June 2006 and January–December 2009. The dataset can be found in the web link http://due.esrin.esa.int/files/Globcover2009_V2.3_Global_zip.

4.2 Application of Numerical Methods in Data Analysis

In Chap. 2, some numerical methods were discussed extensively with examples. In this section, the focus is to practically resolve issues of the application of numerical methods in data analysis. In this respect there are lots of confusing concept that are found online on the subject matter. It is therefore expedient for reader to understand when a numerical method applies to data analysis.

Four concepts in the application of numerical methods in data analysis are presented and discussed.

- First, when the data is used as variables in the formula. This method is fondly used in optimization processes where the modeller needs to see the extreme value of an experiment and the corresponding trends expected at the extreme points. Secondly, this method is used in comparing two or more models with the data as the variable (see Fig. 4.5). Figure 4.5 presents an optimization process via the comparative analysis of Bessel, Chebyshev, Legendre and Lucas polynomials.
  This helps the modeller to comparatively observe the pros and cons of each models and how its significance to the modeller expected outcomes. The third case is when the modeller wishes to test his/her model by plugging-in the dataset into the proposed model. This process helps the modeller to see the initial outcomes of the model, the weakness and strength of the model, and the sections of the model that must be modified. For illustration purpose, let us call this method the 'Data-Into-Formula (DIF)'.
- Second, when the numerical method outcome is plotted side-by-side with the dataset. This technique is used when the modeller wants to compare theoretical standard and the dataset. In this case, the dataset is plotted with theoretical formula or model on the same graph. This process is usually the first validation method in developing a model (see Fig. 4.6). This process is not curve-fitting. In the case of Fig. 4.6, the MISR dataset was plotted alongside the WARSDM model. The model is seen to describe about 60% of the dataset but the trend is not perfect. Hence, this method helps the modeller to see the disparities between is model and real dataset. Secondly, this method is used in ordinary research to show that the dataset that was generated from an experimental process corresponds to the foundational principles.
  This process is very especially when the experimental set-up is new or seemingly modification was done to the experimental set-up. Lastly, this method is used to validate a theoretical research with experimental values. One of elementary ways of showing that the solution of an equation or model is reliable is to plot two plots

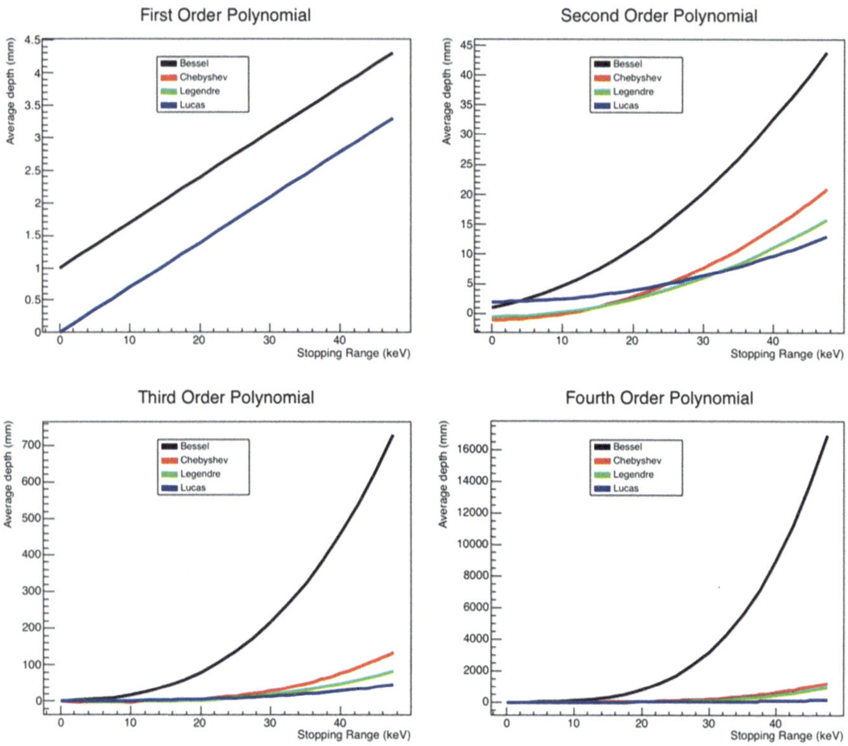

Fig. 4.5 Comparative analysis using two or more techniques or models

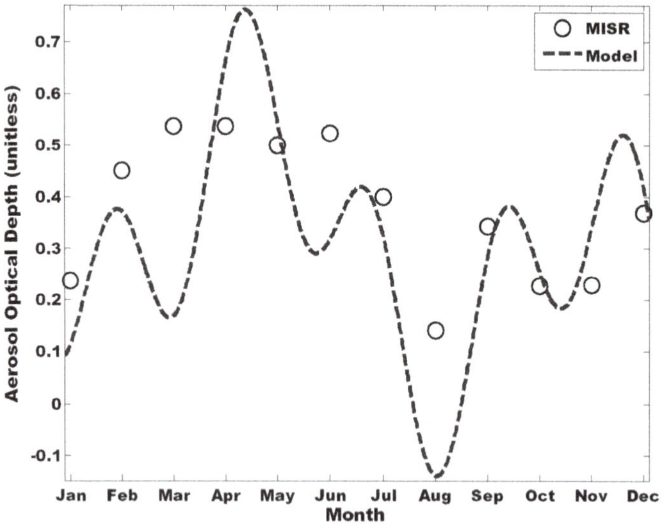

Fig. 4.6 Comparative analysis between a model and dataset

showing the perceived solution and experimental values. For illustration purpose, let us call this method the 'Data-Side-Formula (DSF)'.

- Third, the tracing out of the mathematical or polynomial relationship of an experimental data (maybe using curve-fitting). This method has two concepts i.e. the trivial and complex. In the trivial case, the equations are generated and the regression values are stated alongside. The complex form is the generation of the equation and the application of existing variables of data into the generated equation to see the consistency of the dataset. For example, if I have data 1 and 2, I plot data 1 to get the polynomial relationship say $2x^4 + 3x^2$. Then I see what value of 'x' can replicate data 1. Then I search what value of 'x' that can replicate data 2. If I am able to do that, I have simply got a mathematical relationship that could describe both data 1 and 2 at varying values of 'x'. This method is tedious if it is done manually on Microsoft excel or statistical tool. It is also time consuming. For illustration purpose, let us call this method the 'Data-Derive-Formula (DDF)'.
- Fourth, is using a dataset to obtain the numerical values of unknown terms in a model or equation. For example, in the equation below

$$\psi(\lambda) = a_1^2 \cos\left(\frac{\mathrm{n}_1 \pi\, \tau(\lambda)}{2} x\right) \cos\left(\frac{\mathrm{n}_1 \pi\, \tau(\lambda)}{2} y\right) + \cdots a_n^2 \cos\left(\frac{\mathrm{n}_n \pi\, \tau(\lambda)}{2} x\right) \cos\left(\frac{\mathrm{n}_n \pi\, \tau(\lambda)}{2} y\right)$$

If $\psi(\lambda)$ and $\tau(\lambda)$ is known, then a_1, n_1, x, and y can be obtained via curve-fitting of the plot. This method quite easy but user must be careful not to misplace the x and y coordinates. For illustration purpose, let us call this method the 'Data-Fit-Formula (DFF)'.

The DIF, DSF, DDF and DFF are all scenarios that we shall illustrate using examples in this section.

4.2.1 *Worked Examples on 'DIF' Cases*

In this section, the Microsoft excel program shall be used to solve some of the exercise because majority of user are familiar with the Microsoft excel. The expert user or reader may choose whatsoever computer program they wish. There are three scenarios in this section as discussed in the introductory part of this chapter.

Example 4.1 Plot $y = x^3 + 2x^2 - 5x + 28$ using the dataset presented below.

No.	x
1	7.02
2	7.54
3	7.89
4	9.15
5	9.65
6	9.89
7	11.21
8	11.55

i. Copy the 'x' column into the excel sheet.
ii. Type in the function you want to plot in row one of column "B" and then press the "Enter" key to calculate the value of y in the first cell of the B column. You see the results as shown in the figure below

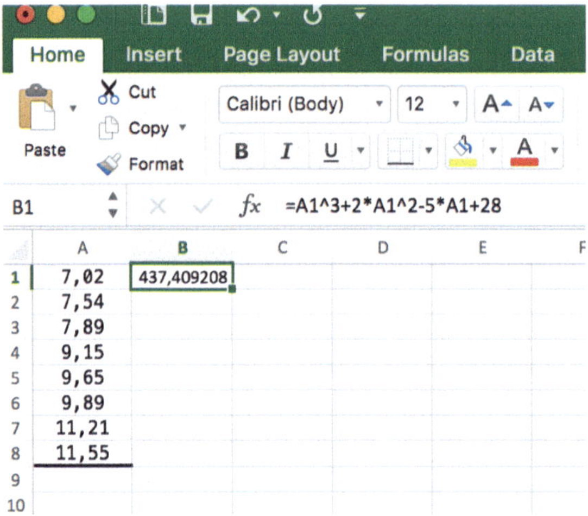

iii. Drag the second column down to the last value in the column (see the figure below)

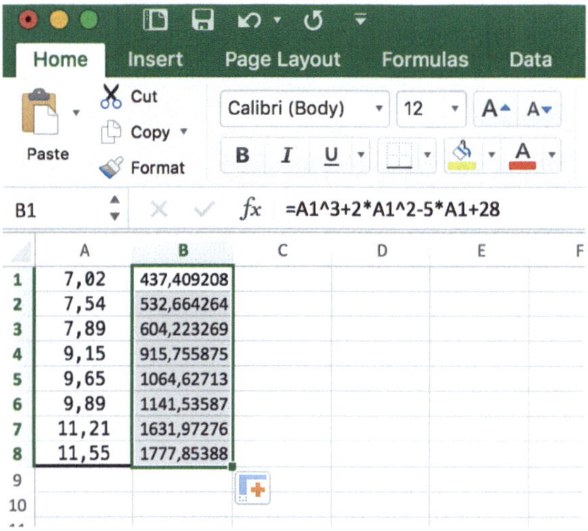

iv. Select the data you want to plot in the chart.
v. Click the Insert tab, and then click X Y Scatter, and under Scatter, pick a chart.

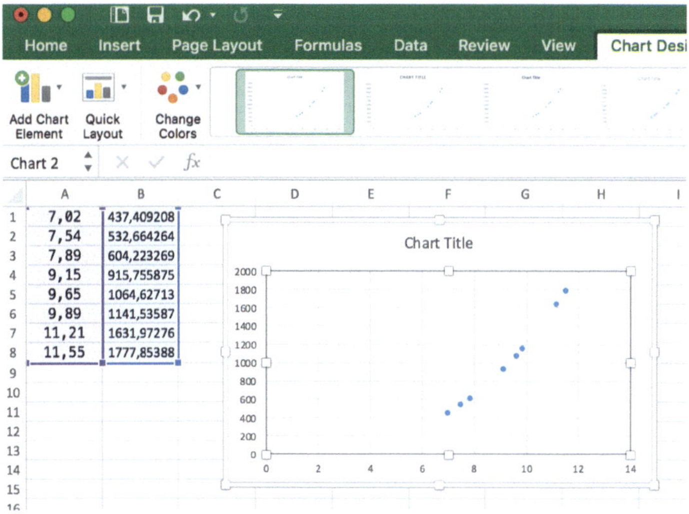

vi. With the chart selected, click the Chart Design tab to do any of the following:
vii. Click Add Chart Element to modify details like the title, labels, and the legend.

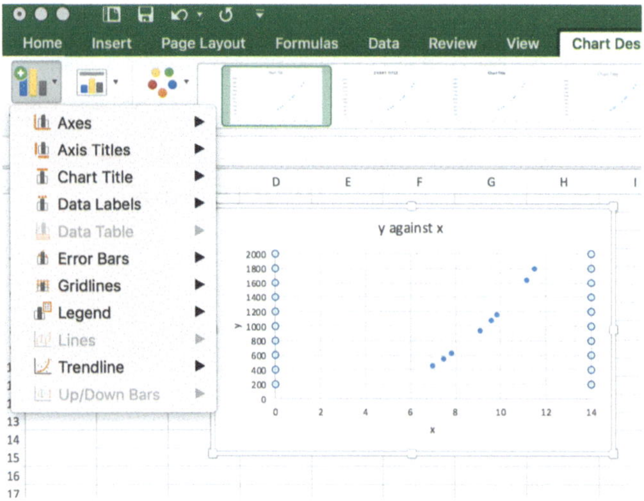

viii. Click Quick Layout to choose from predefined sets of chart elements.
ix. Click one of the previews in the style gallery to change the layout or style.
x. Click Switch Row/Column or Select Data to change the data view.
xi. With the chart selected, click the Design tab to optionally change the shape fill, outline, or effects of chart elements.

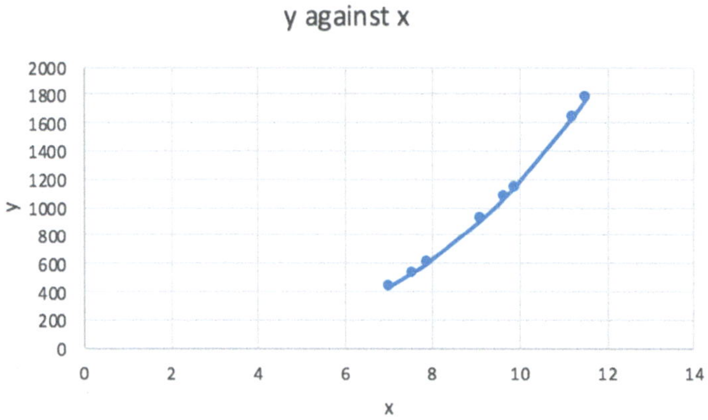

Example 4.2 In this example we consider two variables.

Using $z = 2y^2x^3 - 12yx^2 - 15yx - 128$ and the dataset presented below, plot (a) z against x, (b) z against y.

No.	x	y
1	7.02	1.21
2	7.54	1.25
3	7.89	3.61
4	9.15	5.12
5	9.65	6.31
6	9.89	7.11
7	11.21	7.89
8	11.55	8.11

i. Copy the 'x' and 'y' into column A and B respectively on the excel sheet.
ii. Type in the function you want to plot in row one of column "C" and then press the "Enter" key to calculate the value of z in the first cell of the C column. You see the results as shown in the figure below

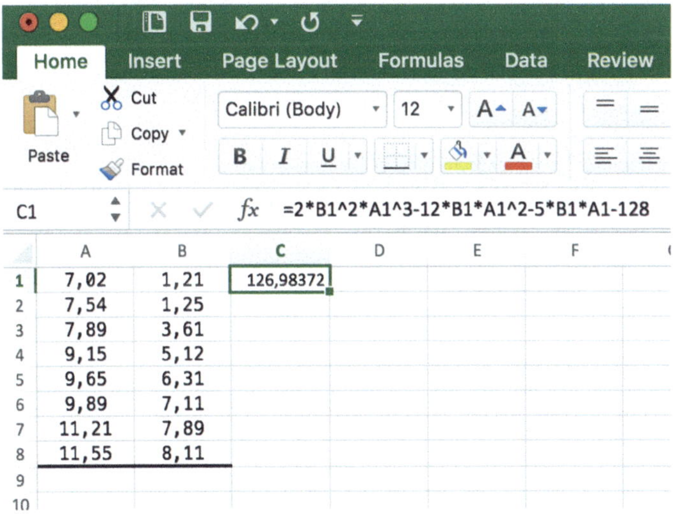

iii. Drag the third column down to the last value in the column (see the figure below)

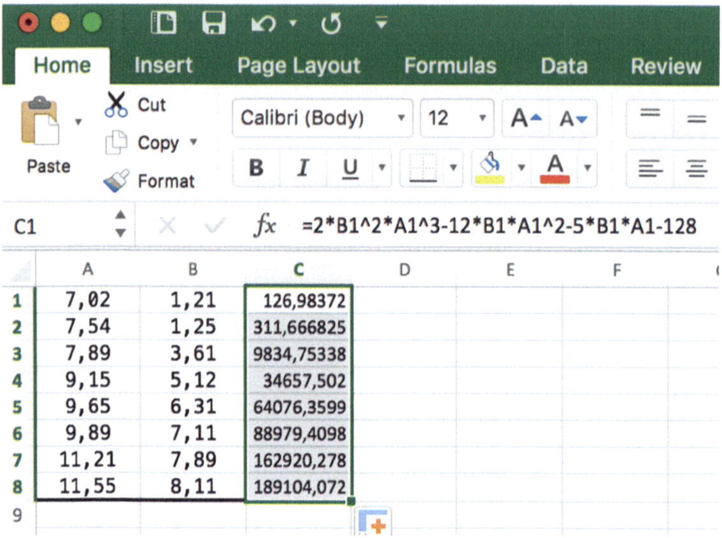

iv. Select the data you want to plot in the chart.

v. Highlight Column B and C, click the Insert tab, and then click X Y Scatter, and under Scatter, pick a chart.

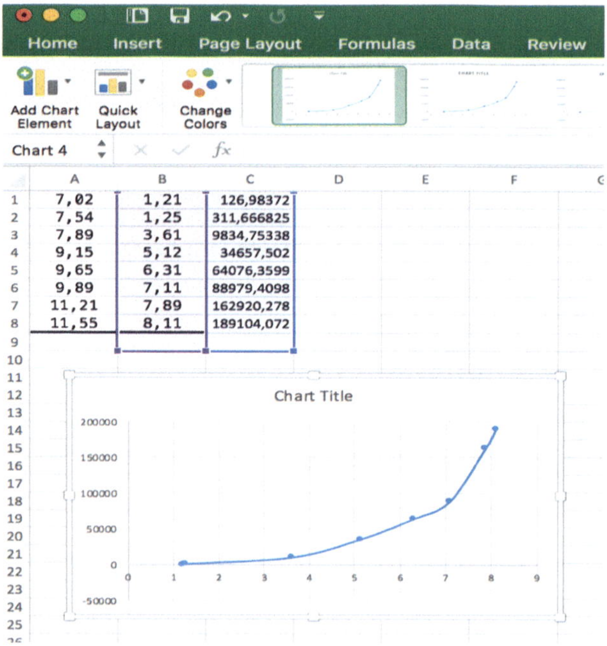

vi. Highlight Column A and C, click the Insert tab, and then click X Y Scatter, and under Scatter, pick a chart.

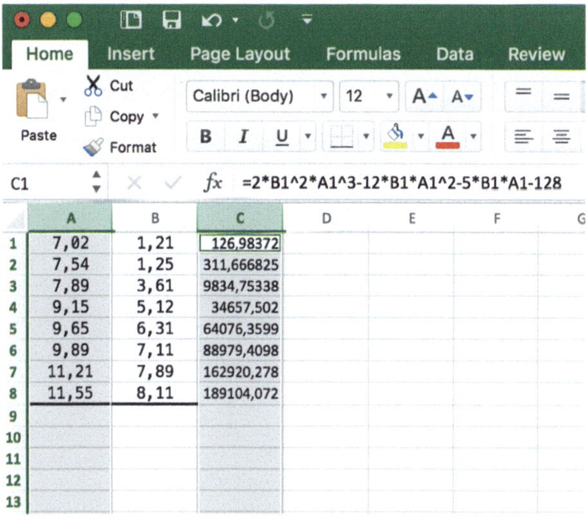

vii. With the chart selected, click the Chart Design tab to do any of the following:

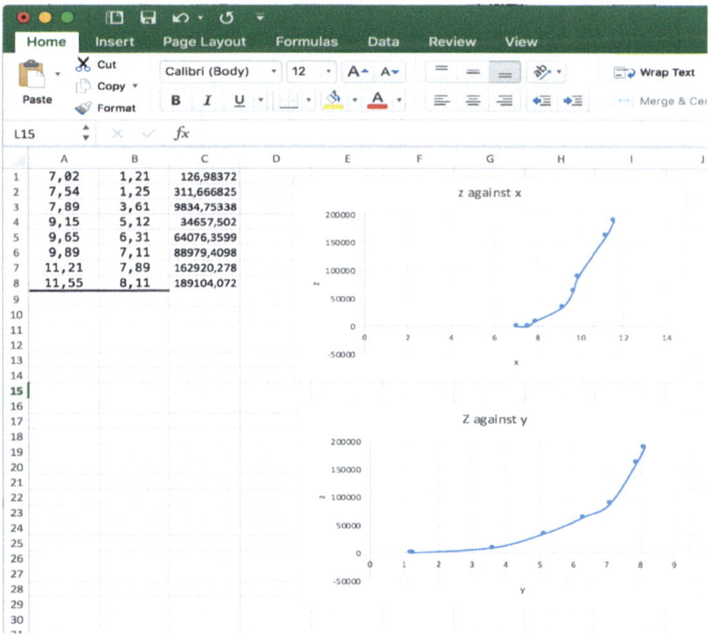

viii. Click Quick Layout to choose from predefined sets of chart elements.
ix. Click one of the previews in the style gallery to change the layout or style.
x. Click Switch Row/Column or Select Data to change the data view.
xi. With the chart selected, click the Design tab to optionally change the shape fill, outline, or effects of chart elements.

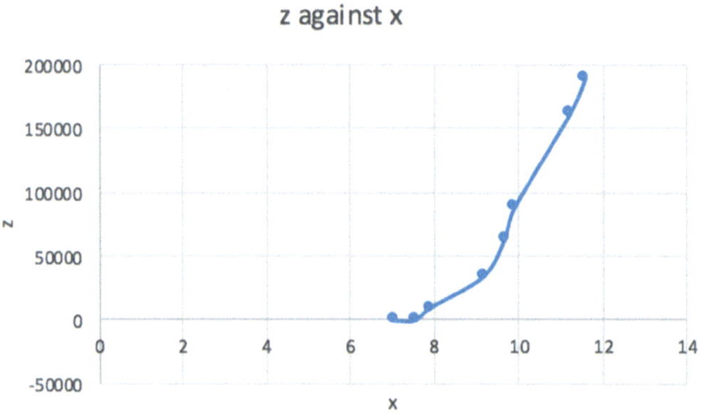

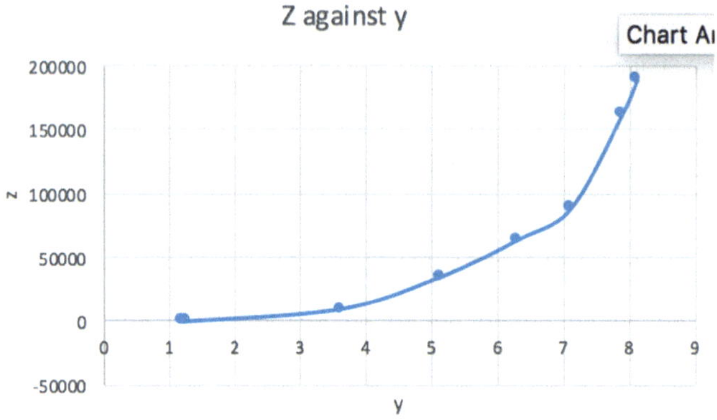

Example 4.3 In this example we consider the second type of 'DIF'.

Using the dataset given below, compare the outcomes between $z = x^3 + 2x^2 - 5x + 28$ and $y = 3x^3 - 2x^2 + 9x - 500$.

i. Enter the second column with $z = x^3 + 2x^2 - 5x + 28$.

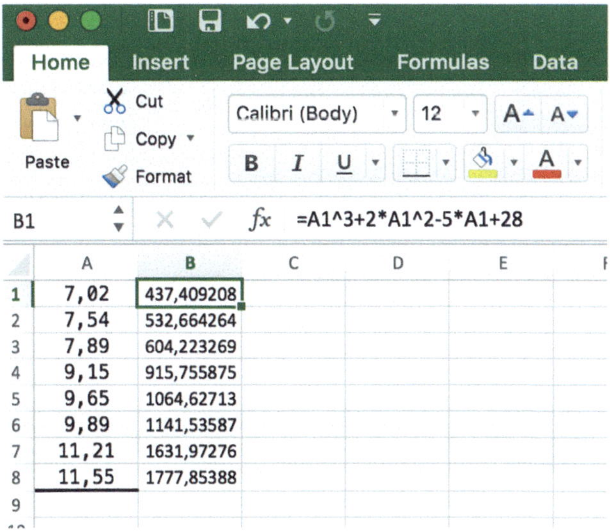

ii. Enter the third column with $y = 3x^3 - 2x^2 + 9x - 500$.

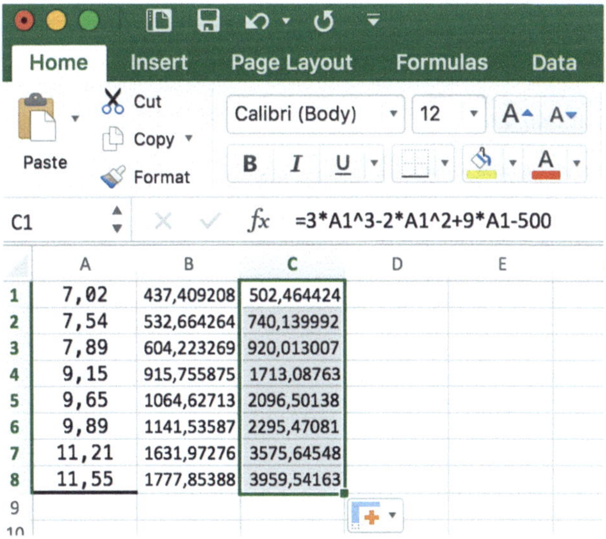

iii. Label the legend.

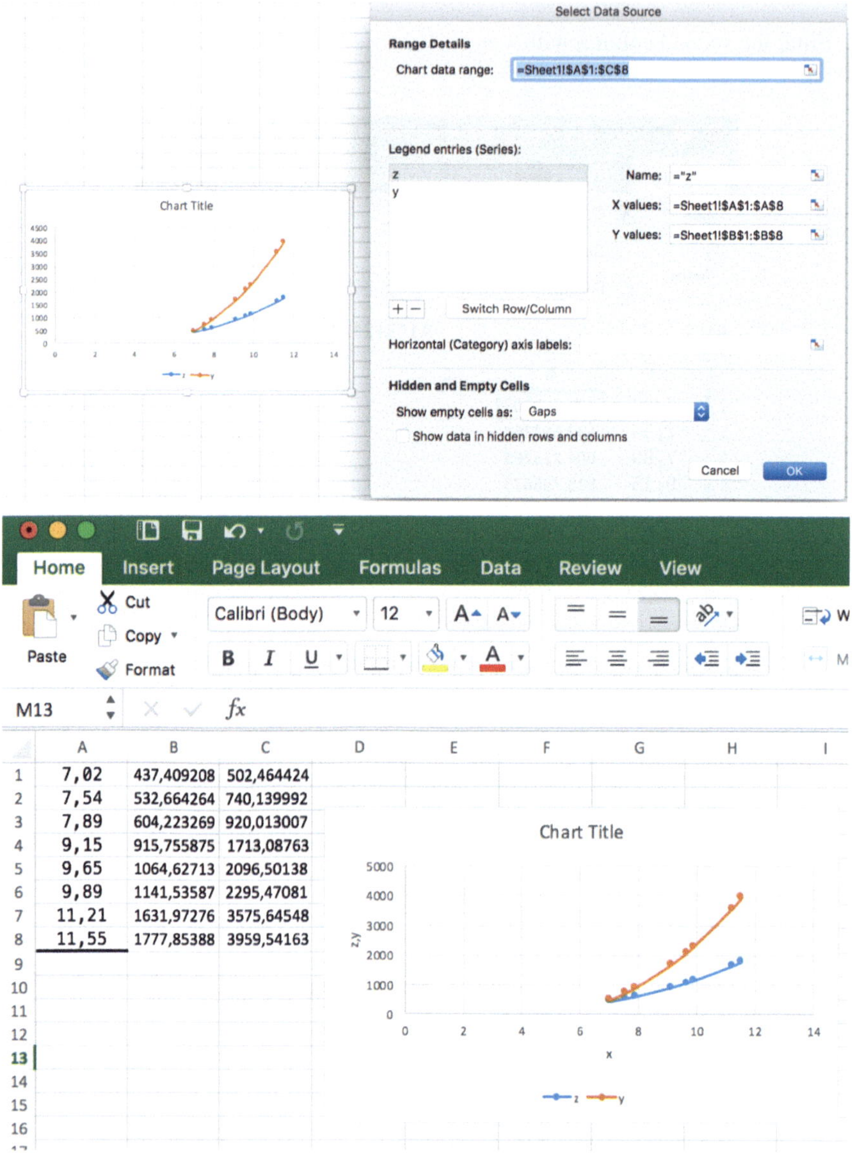

iv. Select all the data you want to graph, click the "Insert" tab, and then select the chart type and sub-type you want to plot. The chart should show a separate plot for the first and second data series on a common Y axis as shown below.

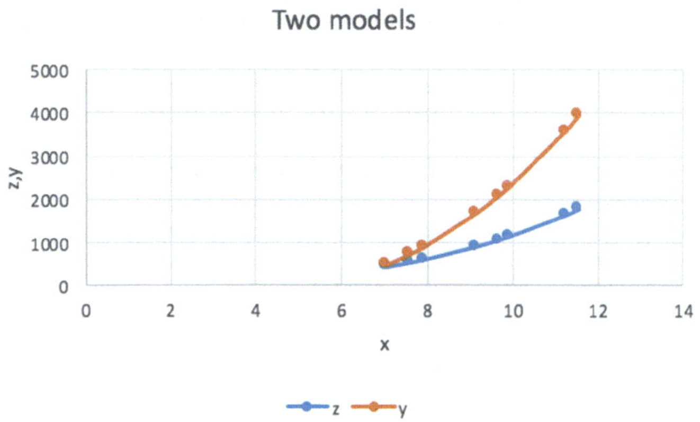

Example 4.4 In this example we consider a three model comparison of 'DIF'.

Using the dataset given below, compare the outcomes between $z = x^3 + 2x^2 - 5x + 28$, $y = 3x^3 - 2x^2 + 9x - 500$ and $R = 2x^3 - 3x^2 + 9x - 69$.

i. Enter the second column with $z = x^3 + 2x^2 - 5x + 28$.

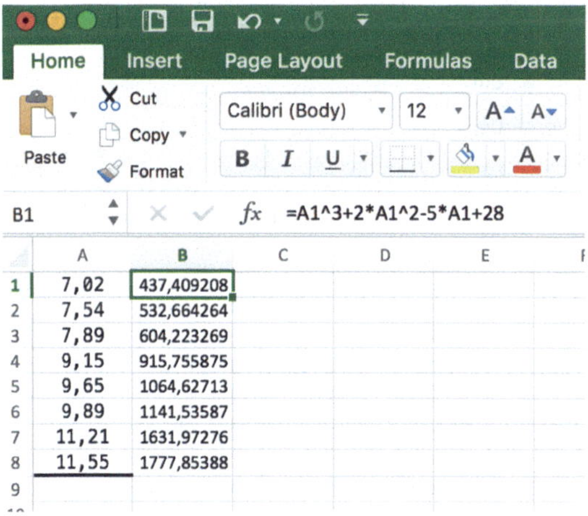

ii. Enter the third column with $y = 3x^3 - 2x^2 + 9x - 500$.

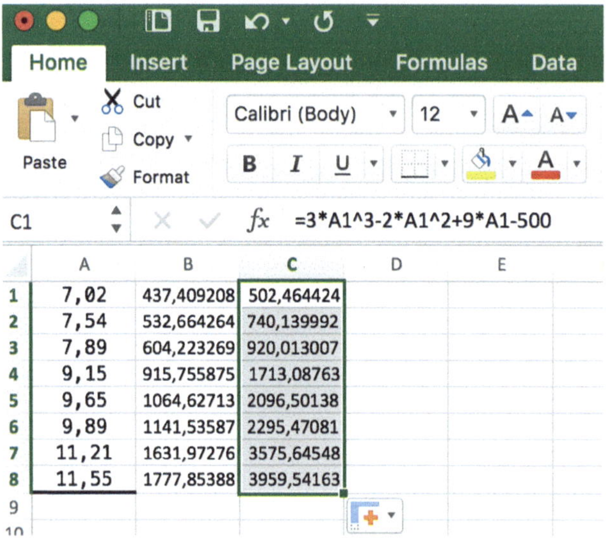

iii. Enter the third column with $y = 3x^3 - 2x^2 + 9x - 500$.

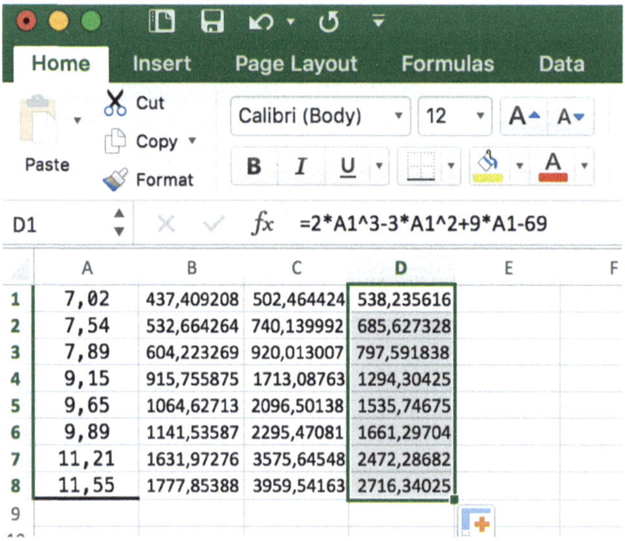

iv. Label the legend.

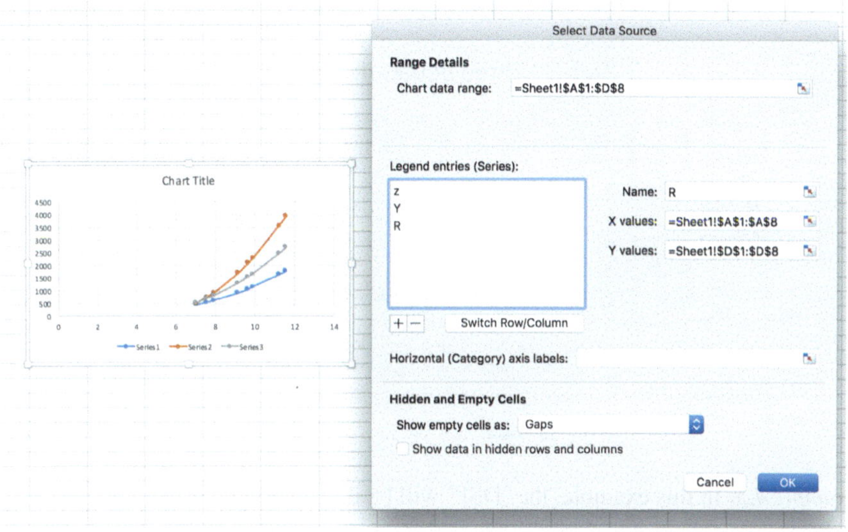

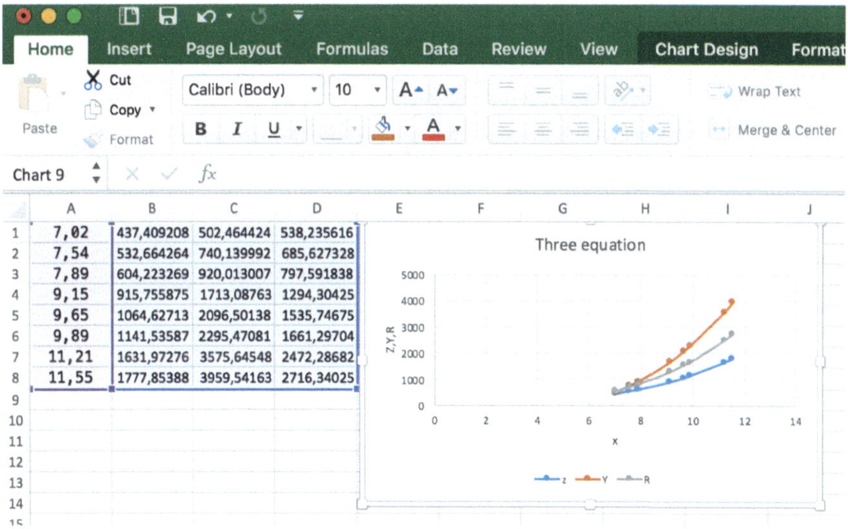

v. Select all the data you want to graph, click the "Insert" tab, and then select the chart type and sub-type you want to plot. The chart should show a separate plot for the first and second data series on a common Y axis as shown below.

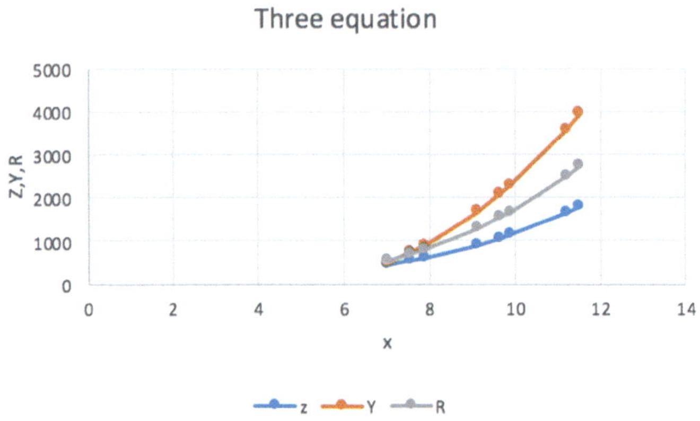

Example 4.5 In this example, the 'DSF' will be illustrated in this example.

In an experiment, the dataset obtained is given as shown in the table below. If theory of the process is mathematical represented as $y = 3t^3 + 2t^2 - 5t - 8$ where t ranges from 1.805 to 1.97. Show the relationship between the experiment and theory.

No.	Y
1	7.02
2	7.54
3	7.89
4	9.15
5	9.65
6	9.89
7	11.21
8	11.55

i. Enter the interval for t in the first column.
ii. Enter the experimental data (Ye) in the second column.
iii. Enter theoretical output (Yt) in the third column.

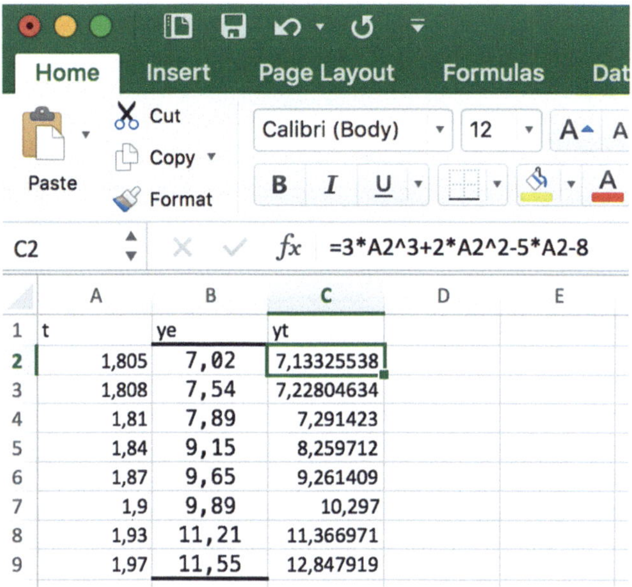

iv. Use the line XY option of insert to plot the two data.

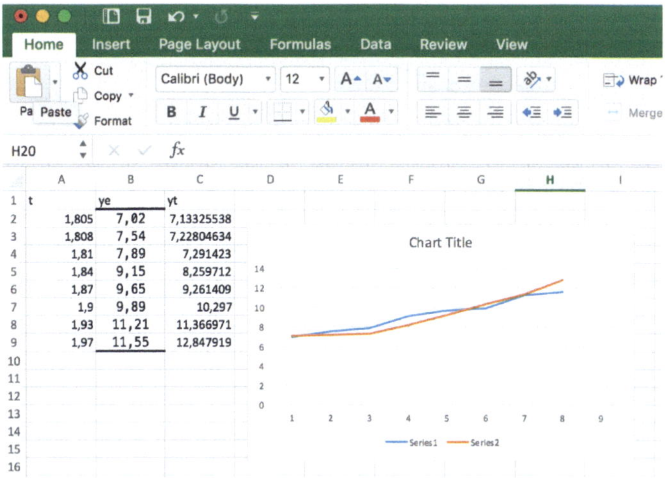

v. Enter the legend and the axis.

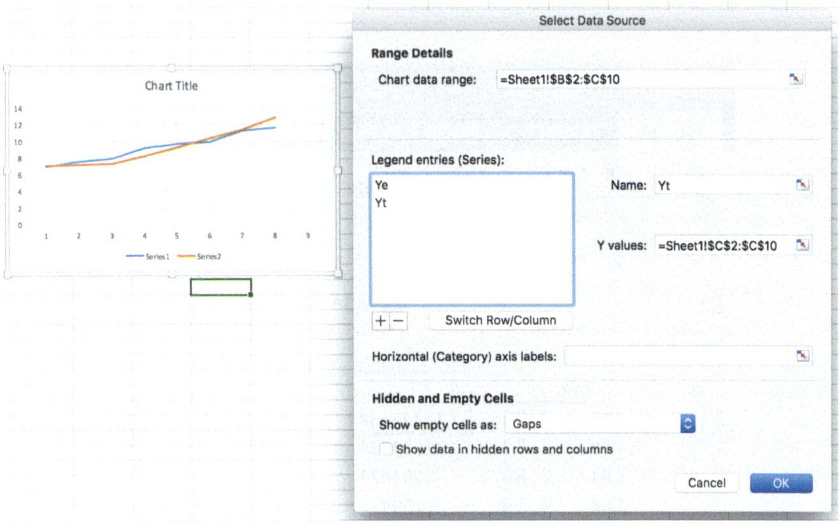

vi. Select all the data you want to graph, click the "Insert" tab, and then select the chart type and sub-type you want to plot. The chart should show a separate plot for the first and second data series on a common Y axis as shown below.

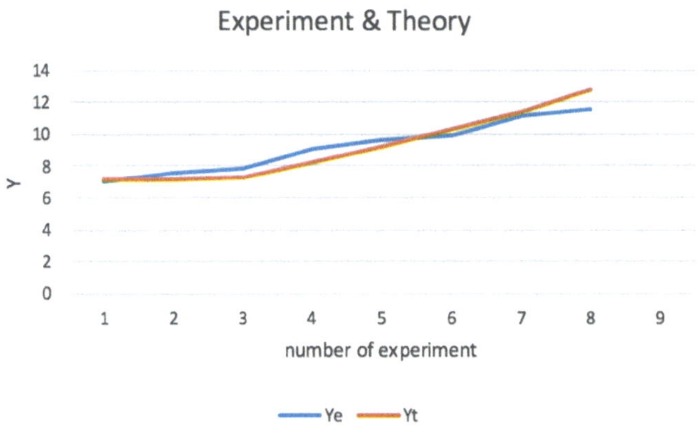

Example 4.6 In this example, the 'DSF' will be illustrated in this example with two theoretical representations.

In an experiment, the dataset obtained is given as shown in the table below. If theory of the process is mathematical represented as $y = 3t^3 + 2t^2 - 5t - 8$ and $z = t^3 + 2t^2 - 4t + 4$ where t ranges from 1.805 to 1.97. Show the relationship between the experiment and theory.

No.	Y
1	7.02
2	7.54
3	7.89
4	9.15
5	9.65
6	9.89
7	11.21
8	11.55

i. Enter the interval for t in the first column.
ii. Enter the experimental data (Ye) in the second column.
iii. Enter theoretical output (Yt) in the third column.

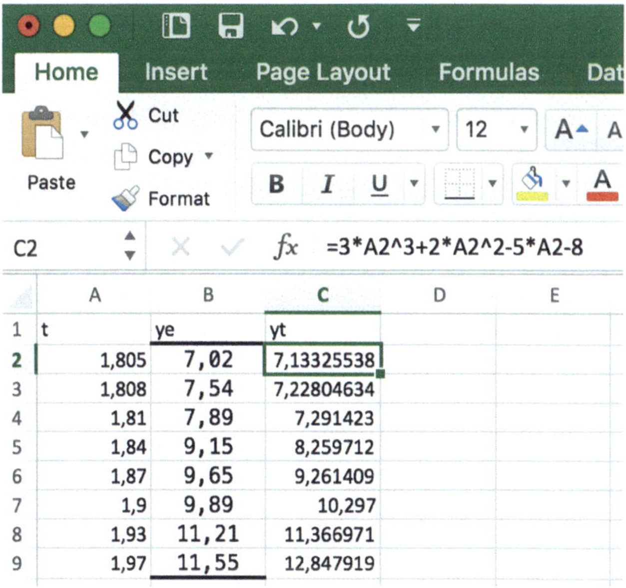

iv. Enter theoretical output (Yt) in the third column.

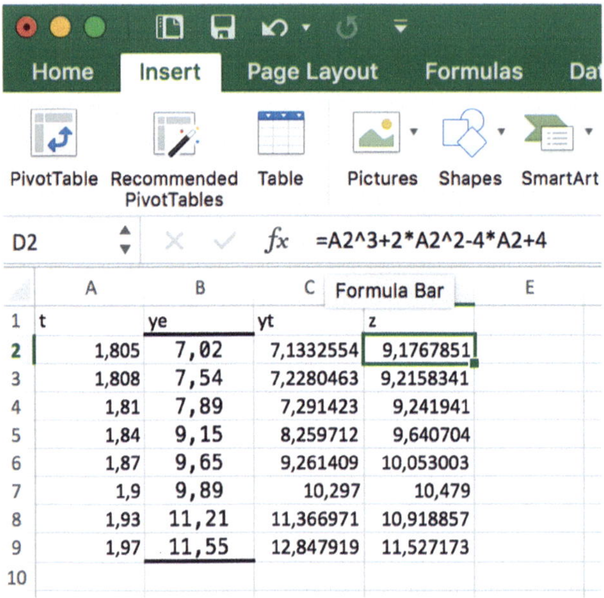

v. Use the line XY option of insert to plot the two data.
vi. Enter the legend and the axis.

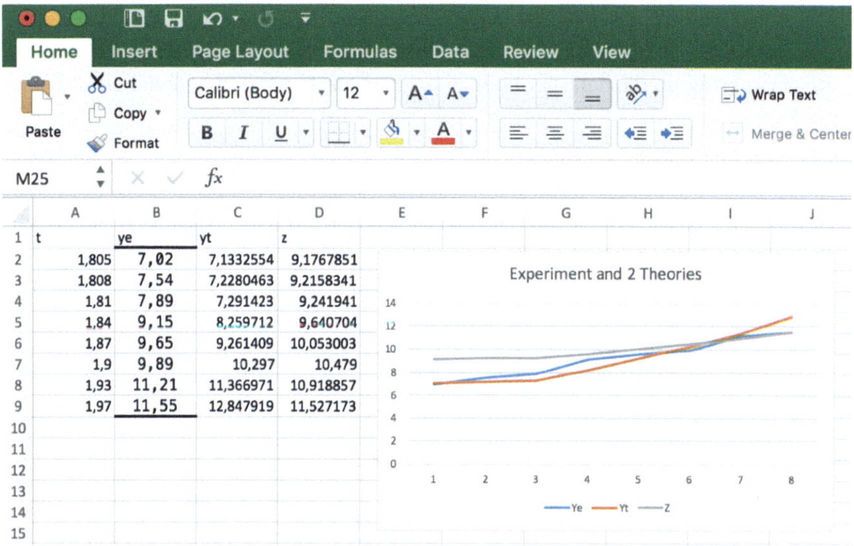

vii. Select all the data you want to graph, click the "Insert" tab, and then select the chart type and sub-type you want to plot. The chart should show a separate plot for the first and second data series on a common Y axis as shown below.

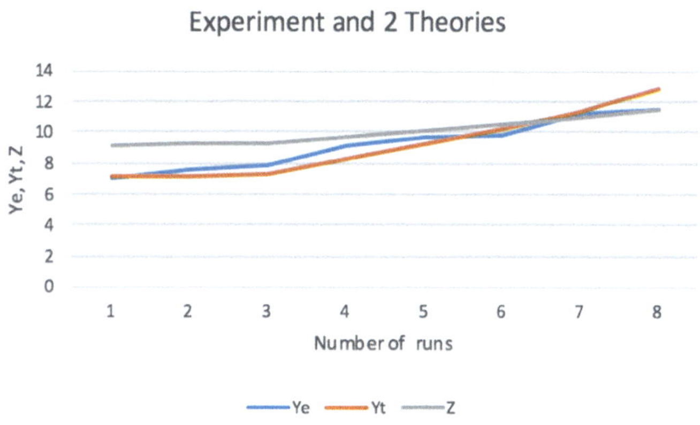

Example 4.7 In this example, the trivial concept of the 'DDF' will be illustrated in this example.

In an experiment, the dataset obtained is given as shown in the table below. Using curve-fitting, show the mathematical expression of the experiment.

No.	Y
1	17.02
2	17.54
3	7.89
4	9.15
5	9.65
6	9.89
7	17.21
8	18.55

i. Enter the interval for Number in the first column.
ii. Enter the experimental data (Ye) in the second column.
iii. Use the scattered plot XY option of insert to plot the two data.

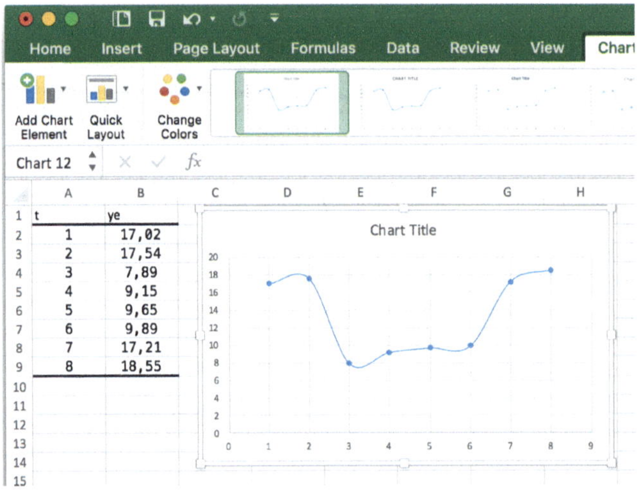

iv. Go to the curve fitting section as shown below

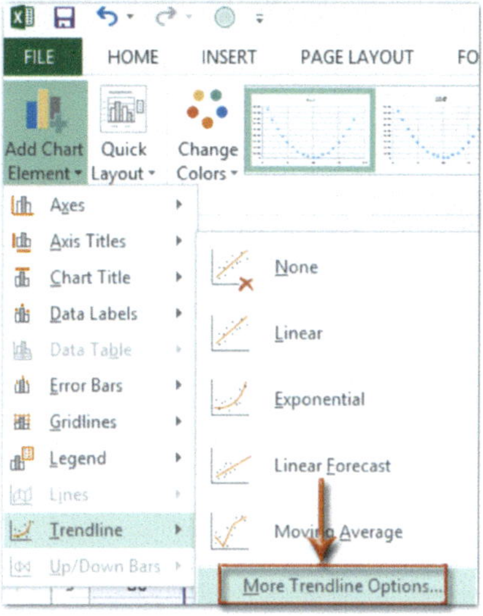

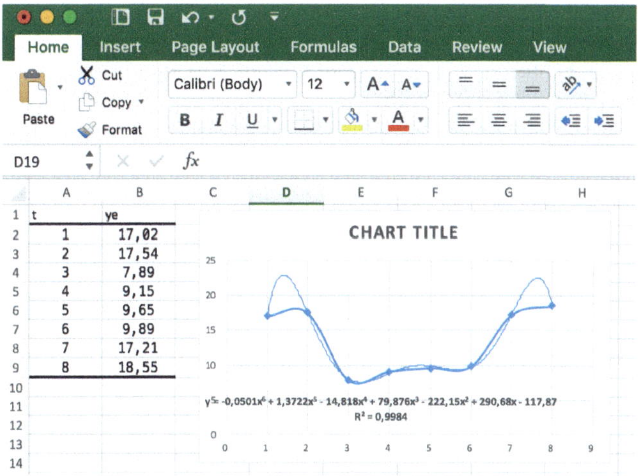

v. Enter the axis title.

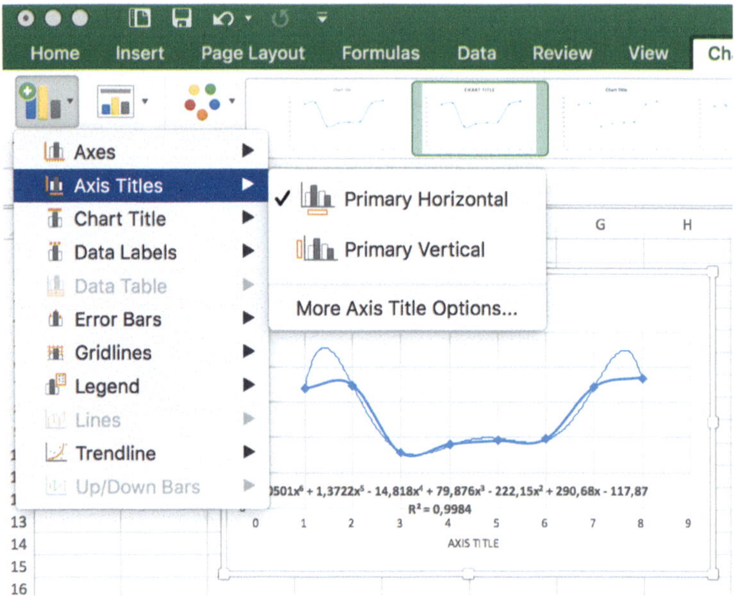

vi. Select all the data you want to graph, click the "Insert" tab, and then select the chart type and sub-type you want to plot. The chart should show a separate plot for the first and second data series on a common Y axis as shown below.

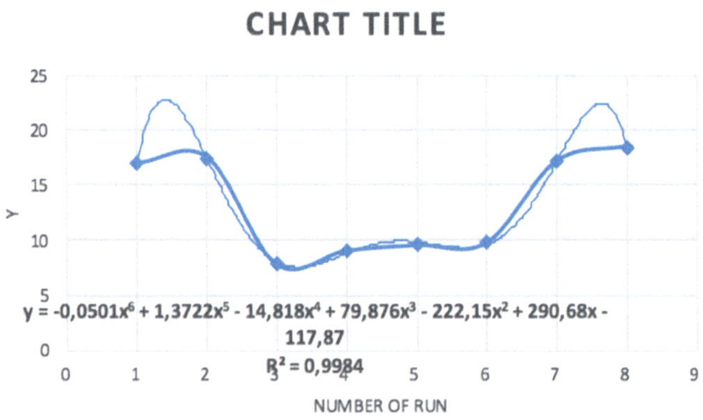

The mathematical expression for the experimental (polynomial) expression is hence

$$y = -0.0501x^6 + 1.3722x^5 - 14.818x^4 + 79.876x^3 + 222.15x^2 - 290.68x - 117.87$$

Example 4.8 In this example, the complex concept of the 'DDF' will be illustrated in this example.

In an experiment, the dataset obtained is given as shown in the table below. Using curve-fitting, show the relationship of the mathematical expression of the experiment

102	85	0	0	17	.	.	0
130	110	0	0	20	.	.	0
404	263	117	0	24	0	0.03	0
477	289	160	0	28	0	0.04	0
569	365	172	0	33	0	0.04	0
625	373	219	0	34	0	0.05	0
769	429	293	0	47	0	0.06	0
925	555	315	0	55	0	0.07	0
867	468	337	0	62	0	0.06	0
909	486	360	0	63	0	0.06	0
799	374	373	0	53	0	0.05	0
737	343	347	0	47	0	0.05	0
633	284	322	0	27	0	0.04	0
643	289	336	0	19	0	0.04	0
593	229	337	0	27	0	0.04	0
626	223	369	0	33	0	0.04	0
539	256	251	0	31	0	0.03	12
674	322	319	0	34	0	0.04	15
656	248	372	0	36	0	0.04	21
625	259	326	0	40	0	0.03	15
919	202	677	0	40	0	0.05	14
996	242	710	0	44	0	0.05	17
740	215	468	0	57	0	0.04	145
808	239	507	0	62	0	0.04	146
824	232	527	0	65	0	0.04	146

(continued)

(continued)

871	226	572	0	73	0	0.04	144
919	227	607	0	85	0	0.04	134
883	183	615	0	85	0	0.04	94
931	239	603	0	89	0	0.04	62
951	242	642	0	67	0	0.04	84
965	230	671	0	64	0	0.04	92
1024	204	759	0	61	0	0.04	114
955	230	665	0	60	0	0.04	118
1030	235	728	0	67	0	0.04	122
838	225	539	0	74	0	0.03	109
1098	220	808	0	70	0	0.04	130
1081	225	783	0	73	0	0.04	124
993	224	708	0	60	0	0.03	127
921	229	632	0	60	0	0.03	127
1051	232	752	0	67	0	0.03	125
1096	254	775	0	67	0	0.03	115
1198	230	906	0	62	0	0.03	111
1110	209	838	0	63	0	0.03	116
808	0	774	0	34	0	0.02	120
768	0	744	0	24	0	0.02	115
707	0	687	0	20	0	0.02	117
447	0	424	0	23	0	0.01	93
579	0	547	0	32	0	0.01	96
632	0	599	0	33	0	0.01	96
518	0	501	0	17	0	0.01	97
455	0	437	0	18	0	0.01	98
377	0	355	0	22	0	0.01	97
222	0	199	0	23	0	0	79
228	0	201	0	27	0	0	82
253	0	217	0	36	0	0	103
269	0	224	0	45	0	0	111
327	0	272	0	55	0	0.01	109
409	0	338	0	71	0	0.01	136
439	0	362	6	71	0	0.01	174
479	0	395	12	72	0	0.01	144

(continued)

(continued)

507	0	439	12	56	0	0.01	13
477	0	402	12	63	0	0.01	13
551	0	471	12	67	0	0.01	126
680	0	606	12	62	0	0.01	124
655	0	599	0	56	0	0.01	119
979	0	918	0	61	0	0.01	129
1274	0	1228	1	45	0	0.02	193

How to determine equations through the curve-fitting technique was illustrated in the last example. The equations of the dataset were derived using Microsoft excel and the replot was done using Mathcad. The data were separated into five parts for easy analysis. Below is the definition of the dataset of the first division. The definition of the x-axis is shown below the immediate dataset.

$$\begin{pmatrix} C \\ D \\ E \\ F \\ G \\ H \\ I \\ J \end{pmatrix} :=$$

102	85	0	0	17 .	.	
130	110	0	0	20 .	.	
404	263	117	0	24	0	0.03
477	289	160	0	28	0	0.04
569	365	172	0	33	0	0.04
625	373	219	0	34	0	0.05
769	429	293	0	47	0	0.06
925	555	315	0	55	0	0.07
867	468	337	0	62	0	0.06
909	486	360	0	63	0	0.06

$$x := \begin{pmatrix} 1 \\ 2 \\ 3 \\ 4 \\ 5 \\ 6 \\ 7 \\ 8 \\ 9 \\ 10 \\ 11 \\ 12 \\ 13 \\ 14 \\ 15 \end{pmatrix}$$

The column was separated with the x and y axis accordingly as shown in the term labelled C, D, E, G and I. After the determination of the equation of the dataset and the regression analysis. The real C and the C1 (determined from the use of new interval is shown in the figure.

C =

	0
0	102
1	130
2	404
3	477
4	569
5	625
6	769
7	925
8	867
9	909
10	799
11	737
12	633
13	643
14	593

D =

	0
0	85
1	110
2	263
3	289
4	365
5	373
6	429
7	555
8	468
9	486
10	374
11	343
12	284
13	289
14	229

E =

	0
0	0
1	0
2	117
3	160
4	172
5	219
6	293
7	315
8	337
9	360
10	373
11	347
12	322
13	336
14	337

G =

	0
0	17
1	20
2	24
3	28
4	33
5	34
6	47
7	55
8	62
9	63
10	53
11	47
12	27
13	19
14	27

I =

	0
0	"."
1	"."
2	0.03
3	0.04
4	0.04
5	0.05
6	0.06
7	0.07
8	0.06
9	0.06
10	0.05
11	0.05
12	0.04
13	0.04
14	0.04

$$C1 := -0.0012 \cdot x^6 + 0.0766 \cdot x^5 - 1.6593 \cdot x^4 + 15.146 \cdot x^3 - 64.987 \cdot x^2 + 247.72 \cdot x - 118.35$$

$$R^2 = 0.9692$$

This exercise is repeated for D and D1, E and E1, G and G1, and H and H1.

C1 =

	0
0	77.945
1	214.136
2	332.205
3	450.824
4	572.387
5	689.176
6	788.663
7	857.957
8	887.374
9	873.15
10	819.291
11	738.556
12	652.575
13	591.108
14	590.438

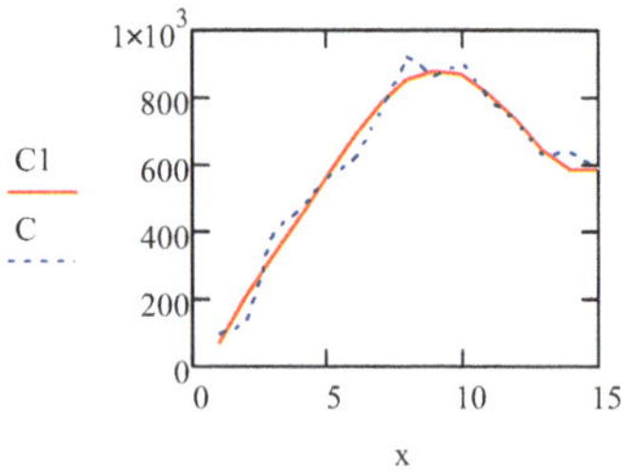

$$D1 := -0.0032 \cdot x^6 + 0.1621 \cdot x^5 - 3.0238 \cdot x^4 + 26.001 \cdot x^3 - 109.08 \cdot x^2 + 275.83 \cdot x - 119.55$$

$$R^2 = 0.9372$$

D1 =

	0
0	70.336
1	160.4
2	220.377
3	281.344
4	349.412
5	417.112
6	472.477
7	505.829
8	514.249
9	503.75
10	489.147
11	491.62
12	533.972
13	633.588
14	793.087

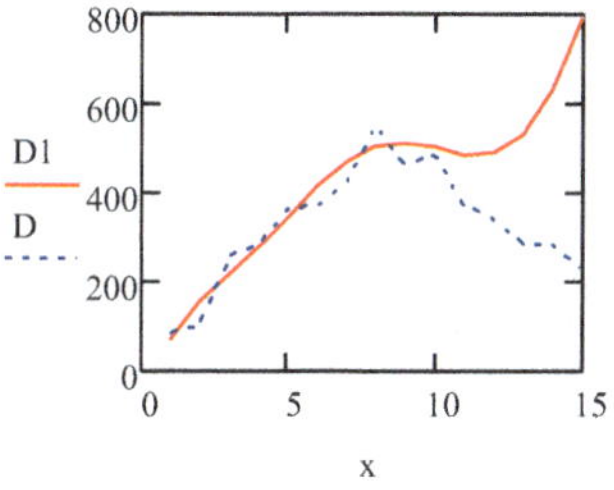

$$E1 := 0.0017 \cdot x^6 - 0.0743 \cdot x^5 + 1.2747 \cdot x^4 - 10.945 \cdot x^3 + 47.122 \cdot x^2 - 40.298 \cdot x - 6.0103$$

$$R^2 = 0.9828$$

E1 =

	0
0	-8.929
1	32.448
2	88.114
3	143.473
4	193.487
5	238.043
6	278.545
7	315.727
8	348.7
9	375.21
10	393.126
11	403.156
12	412.781
13	441.413
14	526.782

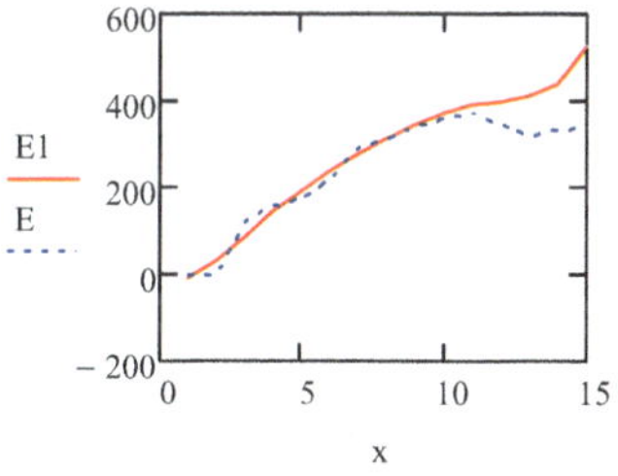

$$G1 := 0.0004 \cdot x^6 - 0.0125 \cdot x^5 + 0.1181 \cdot x^4 - 0.2012 \cdot x^3 - 1.5999 \cdot x^2 + 9.3092 \cdot x + 8.9846$$

$$R^2 = 0.9865$$

G1 =

	0
0	16.599
1	21.109
2	23.901
3	26.818
4	31.383
5	38.304
6	47.272
7	57.045
8	65.819
9	71.887
10	74.59
11	75.551
12	80.2
13	99.584
14	152.47

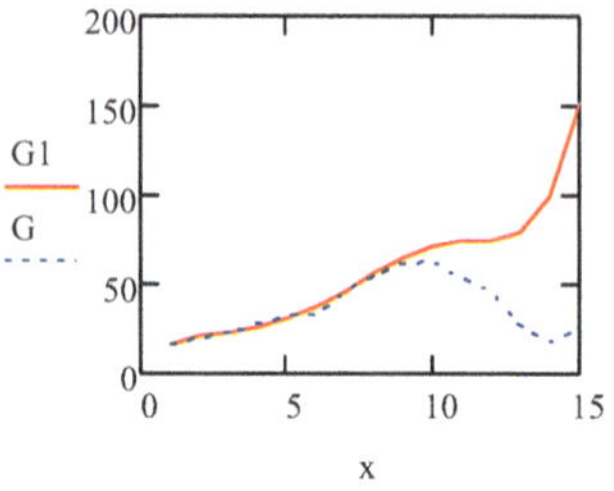

It was observed that there was an error between H and H1.

$$I1 := -1E-06 \cdot x^6 + 4E-05 \cdot x^5 - 0.0006 \cdot x^4 + 0.0041 \cdot x^3 - 0.0119x^2 0.0214 \cdot x + 0.0179$$
$$R^2 = 0.9313$$

I1 =

	0
0	-14.646
1	-547.626
2	$-5.242 \cdot 10^3$
3	$-2.922 \cdot 10^4$
4	$-1.089 \cdot 10^5$
5	$-3.182 \cdot 10^5$
6	$-7.891 \cdot 10^5$
7	$-1.736 \cdot 10^6$
8	$-3.483 \cdot 10^6$
9	$-6.499 \cdot 10^6$
10	$-1.143 \cdot 10^7$
11	$-1.916 \cdot 10^7$
12	$-3.082 \cdot 10^7$
13	$-4.787 \cdot 10^7$
14	$-7.214 \cdot 10^7$

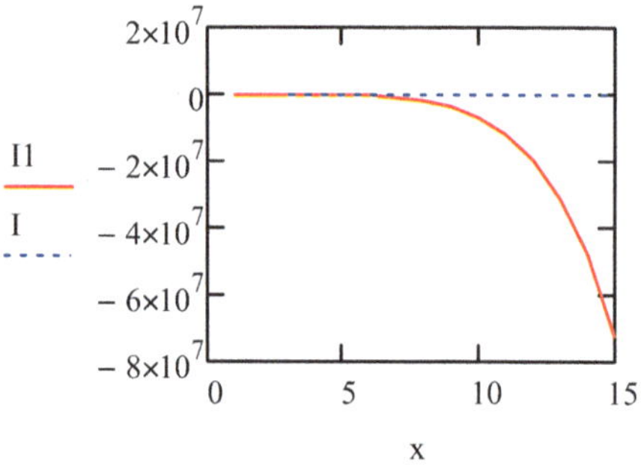

The second set of the dataset is arranged as shown below. The process is repeated as shown above.

$\begin{pmatrix} C \\ D \\ E \\ F \\ G \\ H \\ I \\ J \end{pmatrix} :=$

626	223	369	0	33	0	0.04
539	256	251	0	31	0	0.03
674	322	319	0	34	0	0.04
656	248	372	0	36	0	0.04
625	259	326	0	40	0	0.03
919	202	677	0	40	0	0.05
996	242	710	0	44	0	0.05
740	215	468	0	57	0	0.04
808	239	507	0	62	0	0.04
824	232	527	0	65	0	0.04

$$C2 := 0.0007x^6 - 0.0719x^5 + 2.2362x^4 - 30.515x^3 + 191.88x^2 - 465.89x + 930.64$$
$$R^2 = 0.7388$$

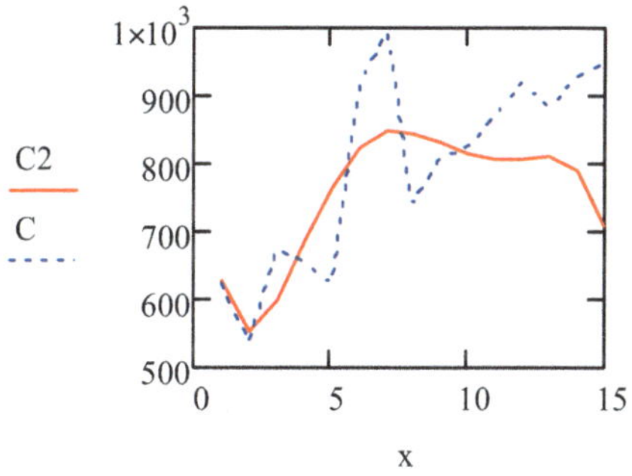

$$D2 := -0.0007 \cdot x^6 + 0.053 \cdot x^5 - 1.3931 \cdot x^4 + 17.273 \cdot x^3 - 105.09 \cdot x^2 + 280.65 \cdot x + 24.631$$
$$R^2 = 0.6042$$

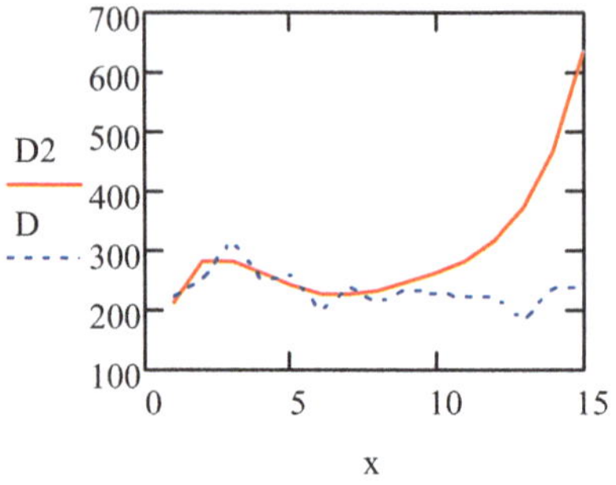

$$E2 := 0.0018x^6 - 0.1398x^5 + 3.8682x^4 - 49.63x^3 + 303.31x^2 - 755.67x + 876.93$$
$$R^2 = 0.7379$$

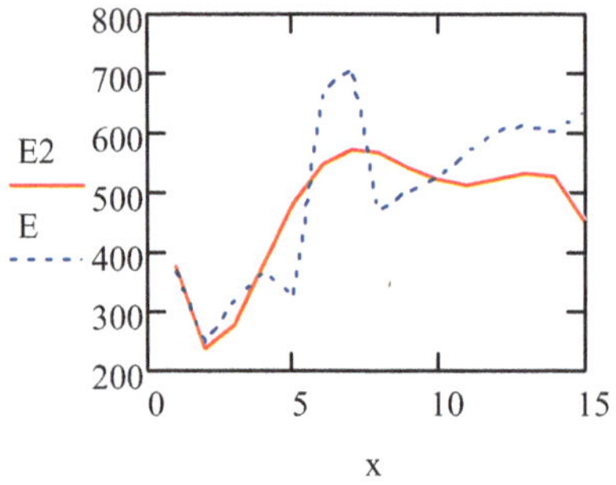

$$G2 := -0.0003 \cdot x^6 + 0.0143 \cdot x^5 - 0.2317 \cdot x^4 + 1.8137 \cdot x^3 - 6.5088 \cdot x^2 + 10.562 \cdot x + 26.672$$

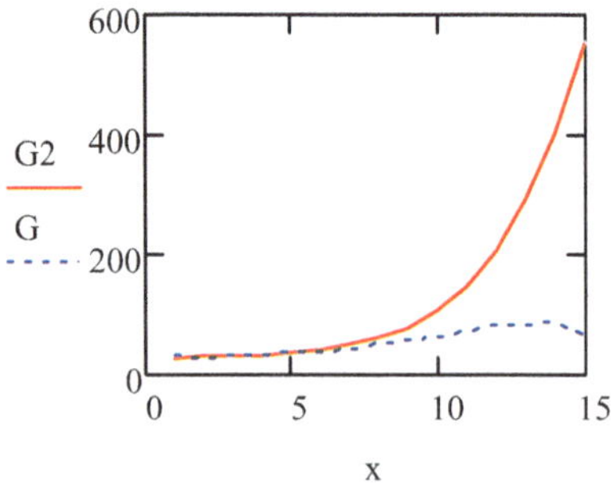

$$I2 := 2E - 08 \cdot x^6 - 3E - 06 \cdot x^5 + 0.0001 \cdot x^4 - 0.0016x^3 + 0.0103x^2 - 0.0273x + 0.057$$
$$R^2 = 0.3281$$

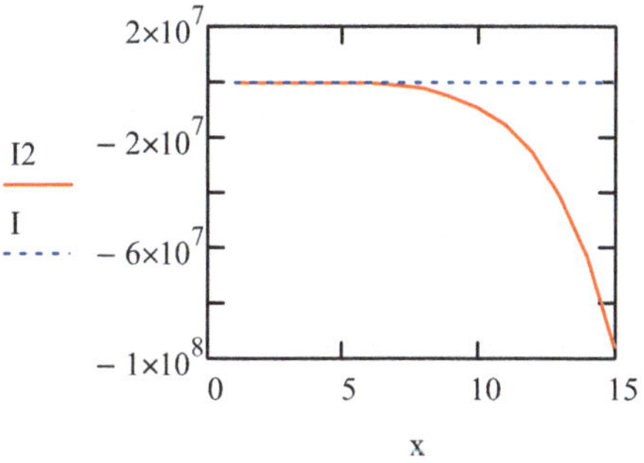

$$J2 := 0.0003 \cdot x^6 + 0.0115 \cdot x^5 - 0.7425 \cdot x^4 + 11.364 \cdot x^3 - 67.185 \cdot x^2 + 159.17 \cdot x - 106.5$$
$$R^2 = 0.9049$$

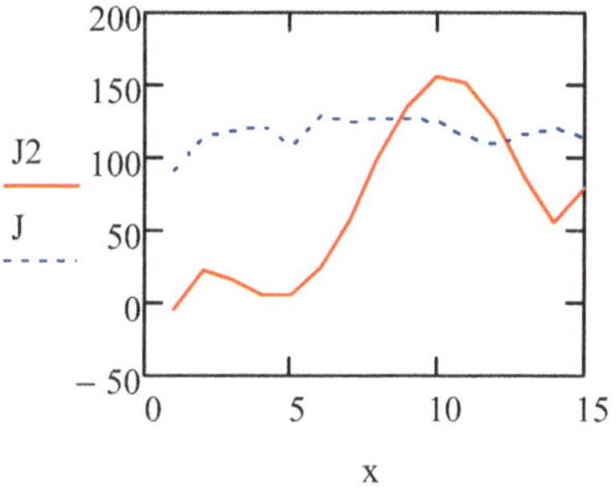

The third set of the dataset is arranged as shown below. The process is repeated as shown above

$$\begin{pmatrix} C \\ D \\ E \\ F \\ G \\ H \\ I \\ J \end{pmatrix} :=$$

965	230	671	0	64	0	0.04
1024	204	759	0	61	0	0.04
955	230	665	0	60	0	0.04
1030	235	728	0	67	0	0.04
838	225	539	0	74	0	0.03
1098	220	808	0	70	0	0.04
1081	225	783	0	73	0	0.04
993	224	708	0	60	0	0.03
921	229	632	0	60	0	0.03
1051	232	752	0	67	0	0.03

$$C3 := 0.006 \cdot x^6 - 0.2961 \cdot x^5 + 5.4571 \cdot x^4 - 47.331 \cdot x^3 + 198.77 \cdot x^2 - 365.38 \cdot x + 1196.8$$
$$R^2 = 0.5629$$

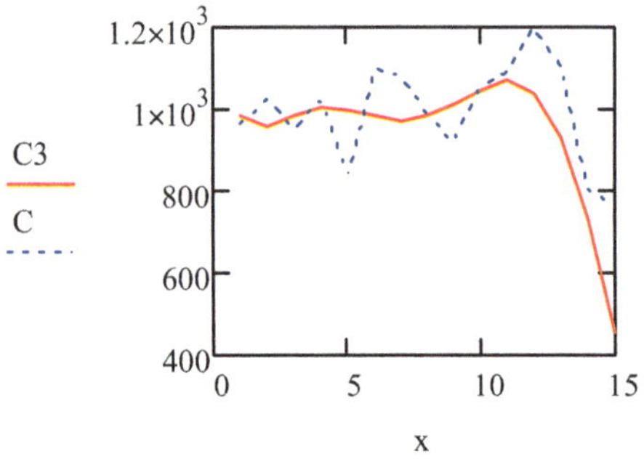

$$D3 := 0.0068 \cdot x^6 - 0.3219 \cdot x^5 + 5.7631 \cdot x^4 - 49.257 \cdot x^3 + 206.1 \cdot x^2 - 382.68 \cdot x + 452.9$$
$$R^2 = 0.9258$$

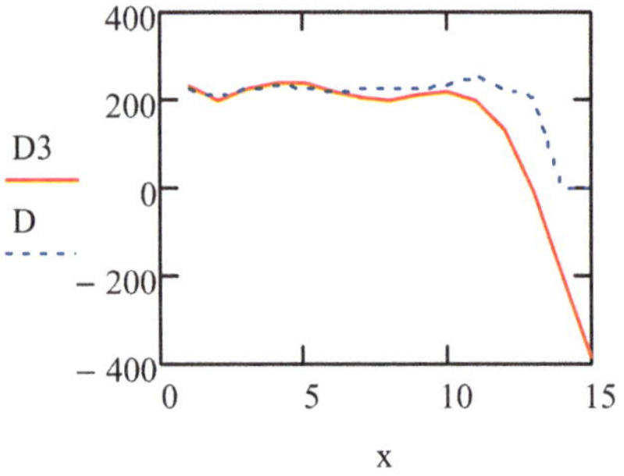

$$E3 := -0.0018 \cdot x^6 + 0.0779 \cdot x^5 - 1.32 \cdot x^4 + 11.424 \cdot x^3 - 50.9 \cdot x^2 + 104.04 \cdot x + 626.78$$
$$R^2 = 0.3716$$

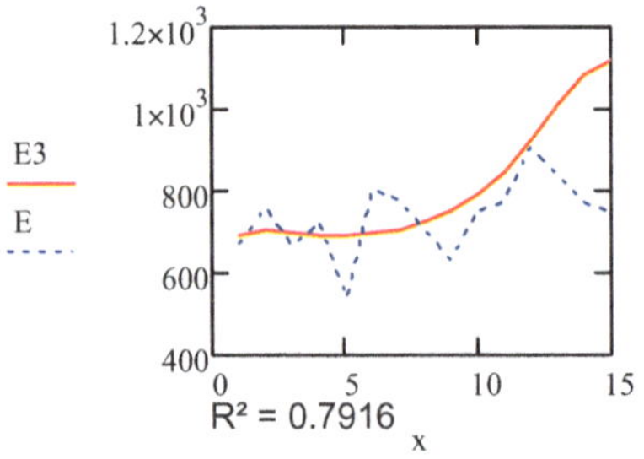

$$I3 := 7E-08 \cdot x^6 - 4E-06 \cdot x^5 + 7E-05 \cdot x^4 - 0.0006 \cdot x^3 + 0.0027 \cdot x^2 - 0.0054 \cdot x + 0.0436$$

$$R^2 = 0.84$$

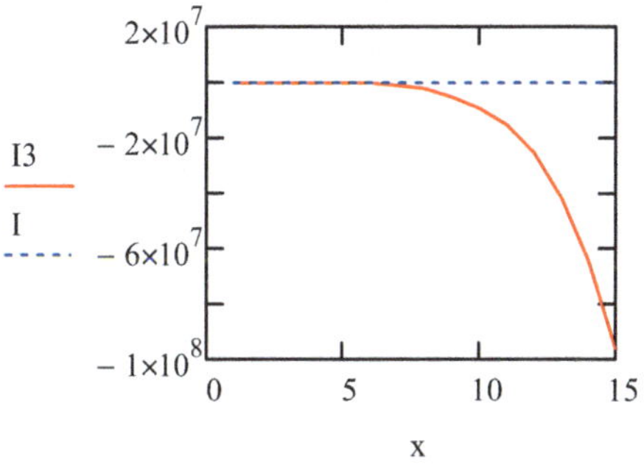

$$R^2 = 0.9206$$

$$G3 := 0.0011 \cdot x^6 - 0.0535 \cdot x^5 + 1.0392 \cdot x^4 - 9.7145 \cdot x^3 + 44.458 \cdot x^2 - 88.311 \cdot x + 117.99$$

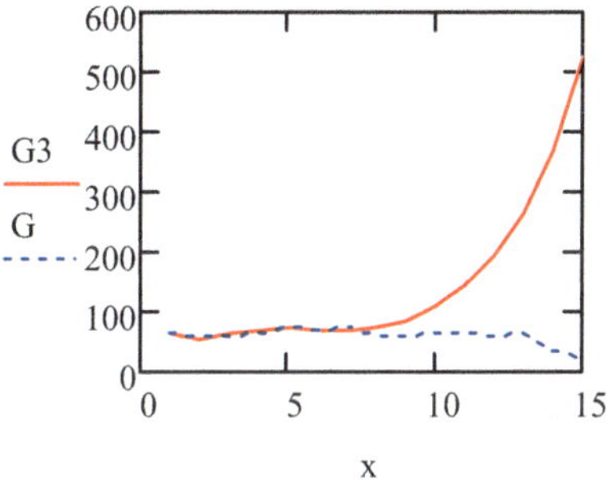

$$J3 := -0.0013 \cdot x^6 + 0.0645 \cdot x^5 - 1.2345 \cdot x^4 + 11.501 \cdot x^3 - 54.195 \cdot x^2 + 122.56 \cdot x + 12.836$$

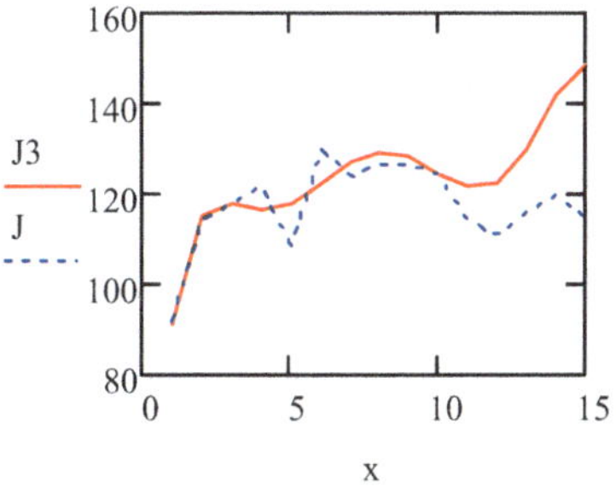

$$\begin{pmatrix} C \\ D \\ E \\ F \\ G \\ H \\ I \\ J \end{pmatrix} :=$$

707	0	687	0	20	0	0.02
447	0	424	0	23	0	0.01
579	0	547	0	32	0	0.01
632	0	599	0	33	0	0.01
518	0	501	0	17	0	0.01
455	0	437	0	18	0	0.01
377	0	355	0	22	0	0.01
222	0	199	0	23	0	0
228	0	201	0	27	0	0
253	0	217	0	36	0	0

The fourth set of the dataset is arranged as shown below. The process is repeated as shown above.

$$C4 := 0.0145 \cdot x^6 - 0.7478 \cdot x^5 + 14.854 \cdot x^4 - 141.24 \cdot x^3 + 652.9 \cdot x^2 - 1360.3 \cdot x + 1525.7$$
$$R^2 = 0.9552$$

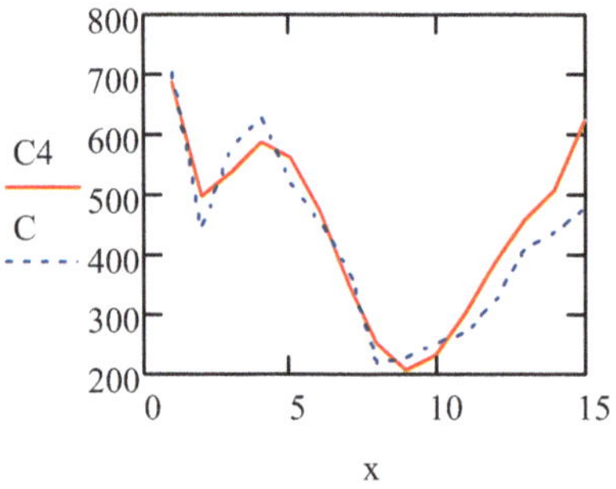

$$F4 := 6E - 14 \cdot x^6 + 0.0005 \cdot x^5 - 0.0162 \cdot x^4 + 0.1843 \cdot x^3 - 0.9145 \cdot x^2 + 1.8846 \cdot x - 1.2$$
$$R^2 = 0.9799$$

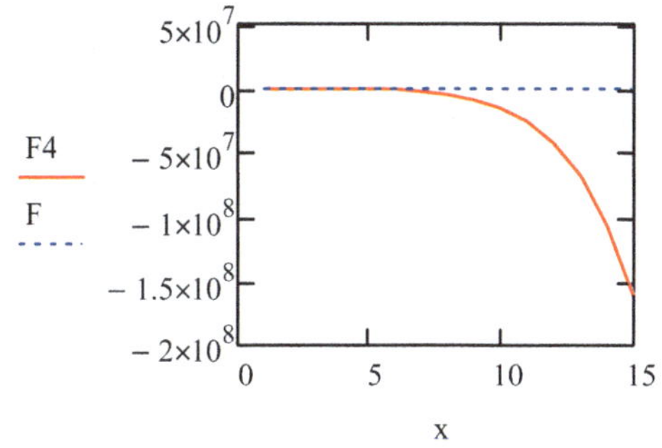

$$G4 := 4E - 05 \cdot x^6 - 0.0019 \cdot x^5 + 0.0103 \cdot x^4 + 0.4769 \cdot x^3 - 6.3716 \cdot x^2 + 24.348 \cdot x + 0.2821$$
$$R^2 = 0.9746$$

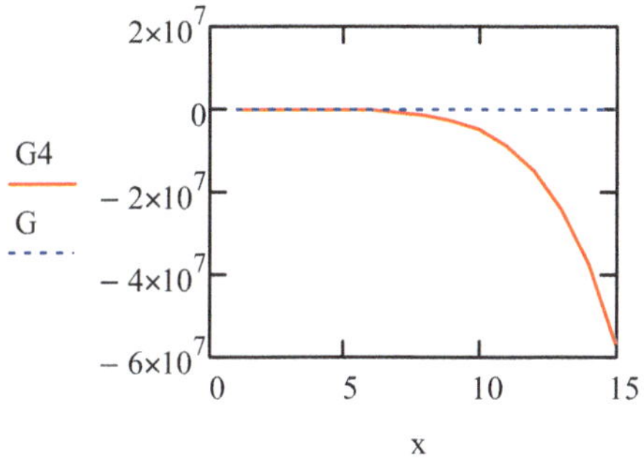

$$E4 := 0.0144x^6 - 0.7464x^5 + 14.86x^4 - 141.9x^3 + 660.19x^2 - 1386.5x + 1526.6$$
$$R^2 = 0.9637$$

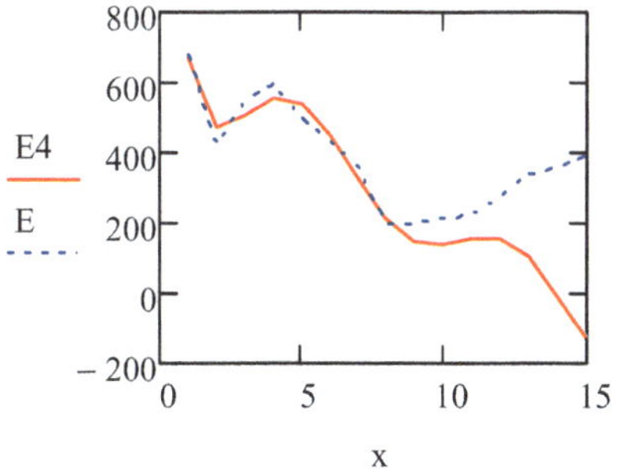

$$I4 := 3E - 07 \cdot x^6 - 2E - 05 \cdot x^5 + 0.0004x^4 - 0.0044x^3 + 0.0229x^2 - 0.0541 \cdot x + 0.0556$$
$$R^2 = 0.8888$$

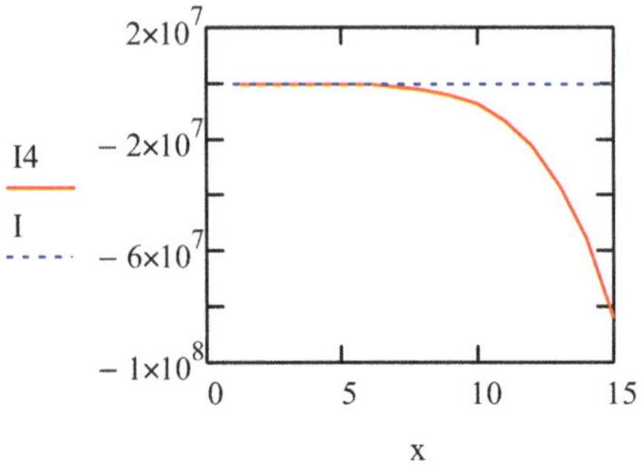

$$J4 := -0.0004x^6 + 0.0098x^5 + 0.0189x^4 - 2.0598x^3 + 18.238x^2 - 59.139x + 158.49$$
$$R^2 = 0.9046$$

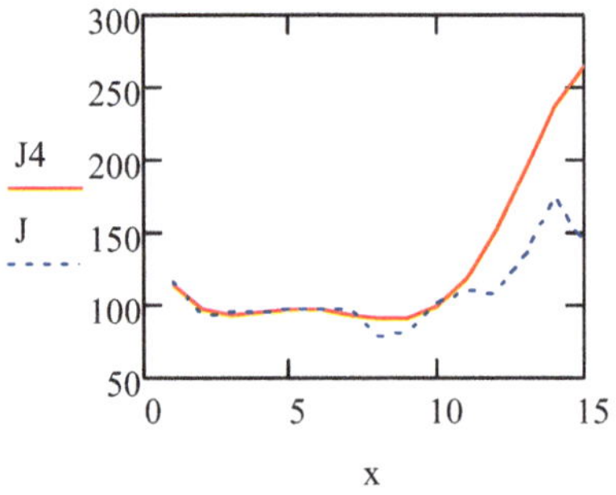

The fifth set of the dataset is arranged as shown below. The process is repeated as shown above

$\begin{pmatrix} C \\ D \\ E \\ F \\ G \\ H \\ I \\ J \end{pmatrix} :=$

507	0	439	12	56	0	0.01	13
477	0	402	12	63	0	0.01	13
551	0	471	12	67	0	0.01	126
680	0	606	12	62	0	0.01	124
655	0	599	0	56	0	0.01	119
979	0	918	0	61	0	0.01	129
1274	0	1228	1	45	0	0.02	193

$$C5 := -3.4236x^6 + 79.163x^5 - 714.8x^4 + 3192.5x^3 - 7327.3x^2 + 8087.8x - 2807$$

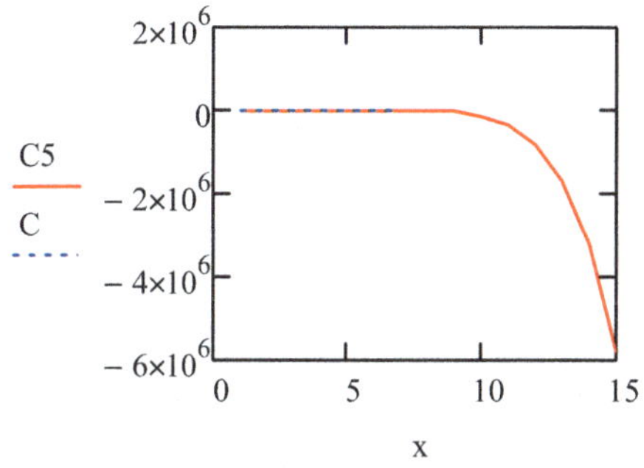

$$F5 := -0.1653x^6 + 3.8708x^5 - 35.424x^4 + 160.48x^3 - 375.91x^2 + 426.15x - 167$$
$$R^2 = 1$$

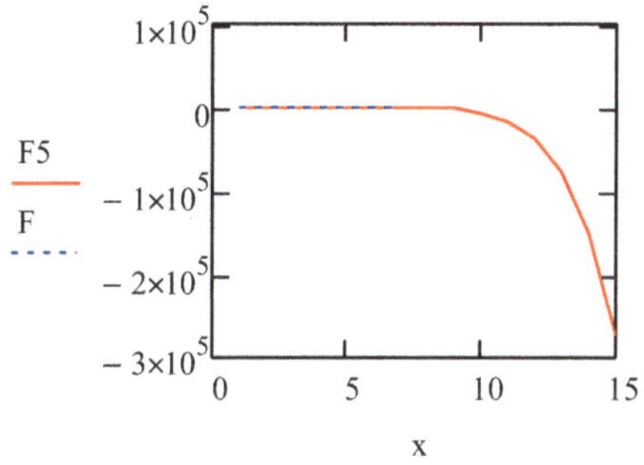

$$G5 := 0.0528 \cdot x^6 + 1.025 \cdot x^5 - 7.4028 \cdot x^4 + 24.875 \cdot x^3 - 42.044 \cdot x^2 + 41.6 \cdot x + 38$$
$$R^2 = 1$$

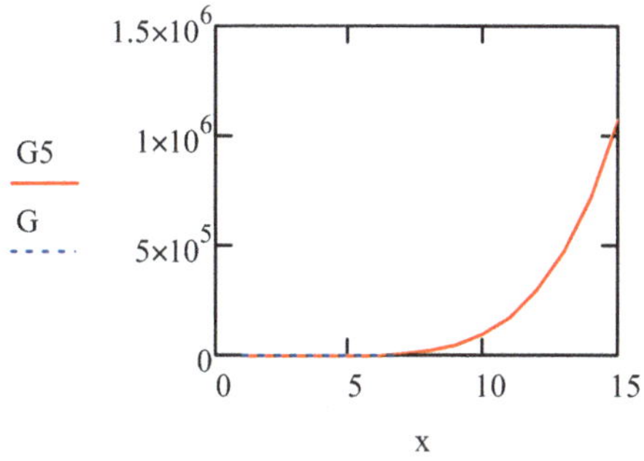

$$E5 := -3.2264 \cdot x^6 + 74.788 \cdot x^5 - 677.12 \cdot x^4 + 3032.6 \cdot x^3 - 6974.2 \cdot x^2 + 7699.2 \cdot x - 2713$$
$$R^2 = 1$$

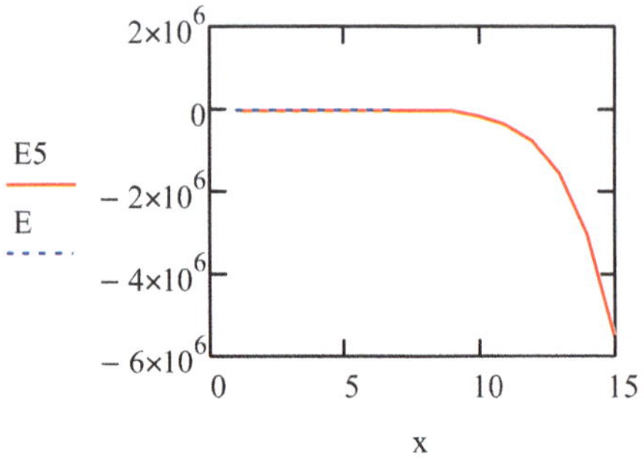

$$J5 := 0.7625 \cdot x^6 - 19.629 \cdot x^5 + 201.85 \cdot x^4 - 1047.5 \cdot x^3 + 2832.4 \cdot x^2 - 3631.9 \cdot x + 1677$$
$$R^2 = 1$$

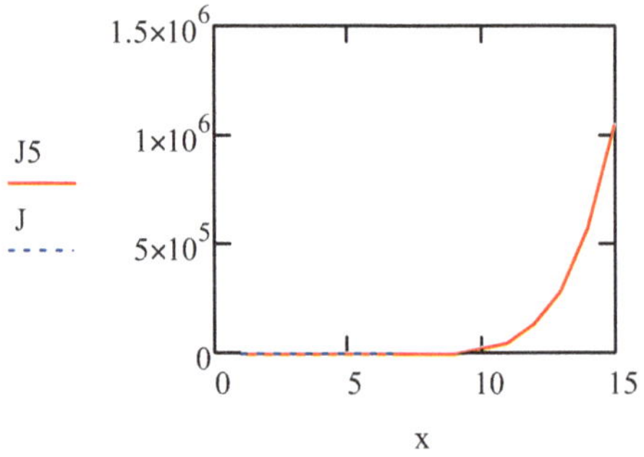

Example 4.9 In this example, the 'DFF' will be illustrated in this example.

If the expression of the dataset is given as $y = 3tx^3 + 2t^2x^{2-5}xt - 8r$. Find the unknown value of x, f, p and r.

x	y
1	7.02
2	7.54
3	7.89
4	9.15

(continued)

(continued)

x	y
5	9.65
6	9.89
7	11.21
8	11.55

The process of obtaining the unknown term in the equation includes

(1) Enter data into spreadsheet.
(2) Plot data on an XY/Scatter plot.
(3) Tell Scatter plot to generate trendline based on data and show the trendline equation on the chart.

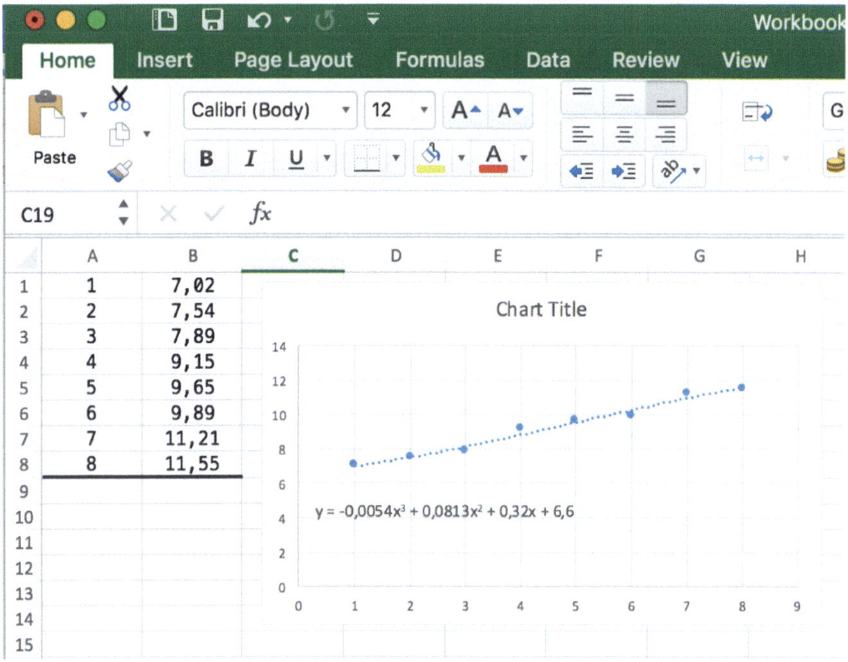

(4) Enter trendline parameters into spreadsheet.

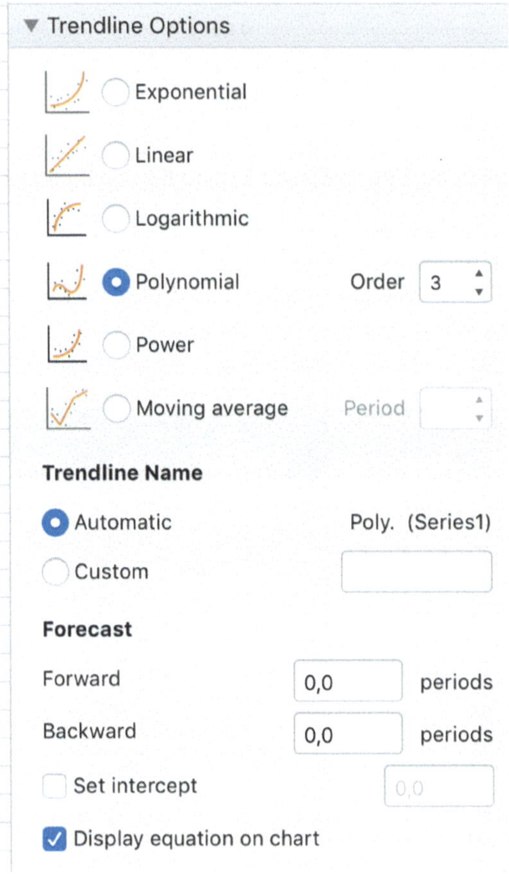

(5) Juxtapose the equation given and generated to know the values of the unknown term. Please note that some software (e.g. Matlab, Origin etc.) can calculate the unknown term directly. We highlight both equations as

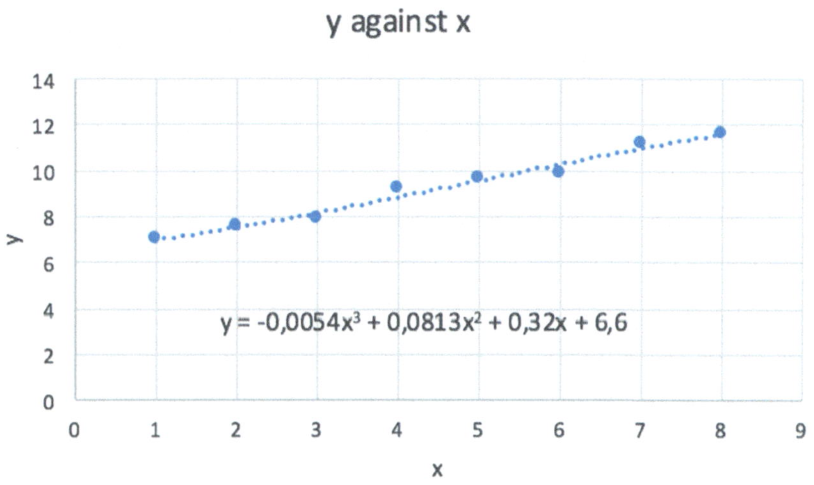

$$y = 3\,tx^3 + 2\,ft^2x^2 - 5\,px - 8r$$
$$y = 0.0054\,x^3 + 0.0813\,x^2 - 0.32\,x - 6.6$$

Hence, the following equations can be extracted

$$3t = 0.0054 \quad (1)$$

$$2ft^2 = 0.0813 \quad (2)$$

$$-5p = -0.32 \quad (3)$$

$$-8r = -6.6 \quad (4)$$

From Eq. (1), $t = 0.0018$
Equation (2) becomes

$$2f(0.0018)^2 = 0.0813$$
$$f = 12546.3$$

From Eq. (3), $p = 0.064$ and from Eq. (4), $r = 0.825$
Hence, the values of the unknown terms are

$$t = 0.0018,\ \ f = 12546.3,\ \ p = 0.064,\ \text{ and } r = 0.825$$

Example 4.10 In this example, the 'DFF' will be illustrated in this example.

If the expression of the dataset is given as $y = 3tx^4 + 3rt^2x^3 + 2rx^2\cos(p\pi) + 5xv^2 + fvp8$. Find the unknown value of x and p.

x	y
1	17.88
2	16.54
3	12.89
4	9.15
5	9.65
6	13.89
7	16.21
8	17.55

The process of obtaining the unknown term in the equation includes

(1) Enter data into spreadsheet
(2) Plot data on an XY/Scatter plot.
(3) Tell Scatter plot to generate trendline based on data and show the trendline equation on the chart.

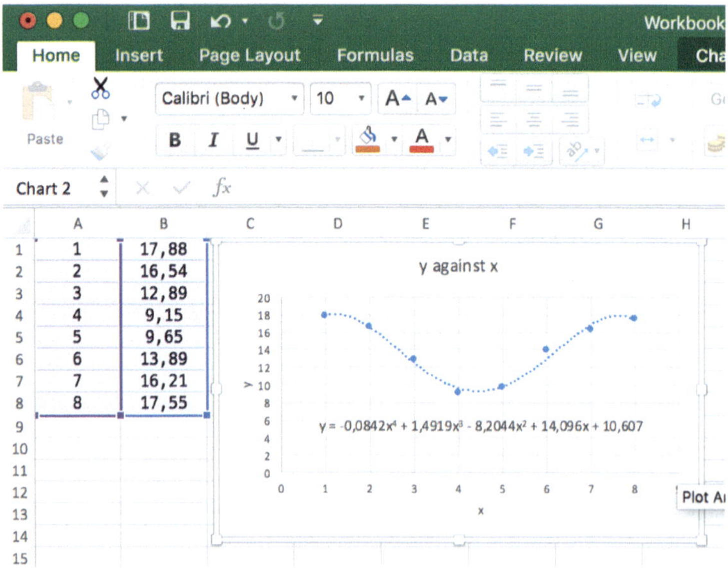

(4) Enter trendline parameters into spreadsheet.

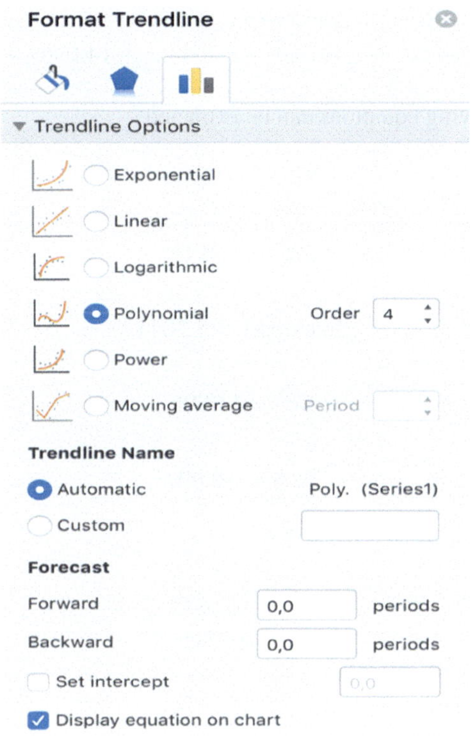

(5) Juxtapose the equation given and generated to know the values of the unknown term. Please note that some software (e.g. Matlab, Origin etc.) can calculate the unknown term directly. We highlight both equations as

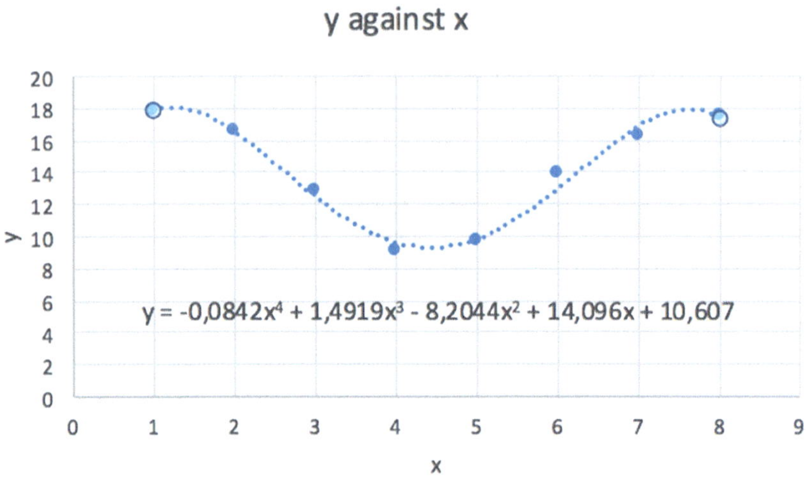

$$y = 3tx^4 + 3r^2x^3 + 2rx^2\cos(p\pi) + 5xv^2 + fvp8$$
$$y = -0.0842x^4 + 1.4919x^3 + 8.2044x^2 + 14.096x + 10.607$$

Hence, the following equations can be extracted

$$3t = -0.0842 \tag{5}$$

$$3rt^2 = 1.4919 \tag{6}$$

$$2r\cos(p\pi) = 8.2044 \tag{7}$$

$$5v^2 = 14.096 \tag{8}$$

$$fvp8 = 10.607 \tag{9}$$

Equation (5) becomes

$$3t = -0.0842$$
$$t = -0.0281$$

Equation (6) becomes

$$3r(-0.0281)^2 = 0.0813$$
$$r = 34.32$$

Equation (7) becomes

$$2{*}34.32{*}\cos(p\pi) = 8.2044$$
$$p\pi = 83.136$$
$$p = 0.4619$$

Equation (8) becomes

$$5v^2 = 14.096$$
$$v^2 = 2.8192$$
$$v = 1.6791$$

Equation (9) becomes

$$\begin{aligned} fvp8 &= 10.607 \\ f^{*}1.67^{*}0.4619^{*}8 &= 10.607 \\ f &= 1.7189 \end{aligned}$$

Hence, the values of the unknown terms are

$$t = -0.0281,\ v = 1.6791,\ f = 1.7189,\ p = 0.4619,\ \text{and}\, r = 34.32$$

The last example is the using of programming language to perform series of numerical manipulations. Using the table below, compare and contrast between images

0.22	0.38	0.21	0.26	0.48	0.40	0.22	0.19
0.28	0.29	0.37	0.46	0.49	0.29	0.26	0.32
0.28	0.39	0.66	0.59	0.50	0.37	0.71	0.36
0.68	0.62	0.61	0.47	0.88	0.44	0.72	0.67
0.68	0.57	0.59	0.85	0.72	0.61	0.66	0.59
0.63	0.64	0.51	0.40	0.75	0.75	0.77	0.58
0.49	0.62	0.39	0.61	0.59	0.65	0.62	0.44
0.57	0.52	0.56	0.34	0.56	0.64	0.37	0.56
0.41	0.38	0.39	0.30	0.35	0.33	0.36	0.42
0.15	0.12	0.28	0.23	0.18	0.27	0.14	0.41
0.28	0.20	0.18	0.34	0.17	0.12	0.34	0.10
0.14	0.14	0.15	0.42	0.21	0.63	0.11	0.21

Using the format of C++ code written below, the various figures will be obtained if the reader is versed in C++.

```
void aerosol(const char *dirname="damatururet.txt", const char *dirnome="damaturueff.txt",
const char *dirnime="damaturusiz.txt")
{
   double a,b,c,d,e,f,g,h,i,j,k,l,m,n;
   double sa,sb,sc,sd,se,sf,sg,sh,si,sj,sk,sl,sm,sn;
      TString dir = gSystem->UnixPathName(gInterpreter->GetCurrentMacroName());
```

```
dir.ReplaceAll("sem.C","");
dir.ReplaceAll("/./","/");
TFile *fe = new TFile("dirname.root","RECREATE"); //create file data.root
TTree *tree = new TTree("tree","data from ascii file");
Double_t nlines = tree->ReadFile(Form(dirname,dir.Data()),"a:b:c:d:e:f:g:h:i:j:k:l:m:n");//create tree with
TString dor = gSystem->UnixPathName(gInterpreter->GetCurrentMacroName());
dor.ReplaceAll("sem.C","");
dor.ReplaceAll("/./","/");
TFile *fo = new TFile("dirnome.root","RECREATE"); //create file data.root
TTree *tree1 = new TTree("tree1","data from ascii file");
Double_t nlinesi = tree1->ReadFile(Form(dirnome,dor.Data()),"sa:sb:sc:sd:se:sf:sg:sh:si:sj:sk:sl:sm:sn");
TString diir = gSystem->UnixPathName(gInterpreter->GetCurrentMacroName());
diir.ReplaceAll("sem.C","");
diir.ReplaceAll("/./","/");
TFile *fi = new TFile("dirnime.root","RECREATE"); //create file data.root
TTree *tree2 = new TTree("tree2","data from ascii file");
Double_t nlinest = tree2->ReadFile(Form(dirnime,diir.Data()),"fa:fb:fc:fd:fe:ff:fg:fh:fi:fj:fk:fl:fm:fn");
{ TCanvas *d = new TCanvas("d", "Face 2",1400,800);
TMultiGraph * mh = new TMultiGraph();
tree->Draw("c:a");
TGraph *gh = new TGraph(tree->GetSelectedRows(),tree->GetV2(), tree->GetV1());
gh->SetName(" ");
gh->Draw();
gh->SetLineColor(2);
gh->SetMarkerStyle(21);
gh->SetMarkerSize(7);
gh->SetLineWidth(6);
mh->Add(gh);
tree->Draw("m:a");
TGraph *gj = new TGraph(tree->GetSelectedRows(),tree->GetV2(), tree->GetV1());
gj->SetName(" ");
gj->Draw();
gj->SetLineColor(4);
gj->SetMarkerStyle(21);
gj->SetLineWidth(6);
gj->SetMarkerSize(7);
mh->Add(gj);
mh->Draw("A");
mh->GetXaxis()->SetTitle("Month of Year");
mh->GetYaxis()->SetTitle("Aerosol Retention");
TLegend *legend1=new TLegend(0.6,0.35,0.88,0.55);
legend1->SetTextFont(50);
legend1->SetTextSize(0.04);
legend1->AddEntry(gh,"Aerosol Retention 2001","l");
legend1->AddEntry(gj,"Aerosol Retention 2012","l");
legend1->Draw();
d->Update();
d->Modified();
```

```
}
{ TCanvas *r = new TCanvas("r", "Face 2",1400,800);
      TMultiGraph * mr = new TMultiGraph();
      tree1->Draw("sc:sa");
      TGraph *gi = new TGraph(tree1->GetSelectedRows(),tree1->GetV2(), tree1->GetV1());
gi->SetName(" ");
gi->Draw();
gi->SetLineColor(3);
gi->SetMarkerStyle(21);
gi->SetLineWidth(6);
gi->SetMarkerSize(7);
      mr->Add(gi);
      tree1->Draw("sm:sa");
      TGraph *gk = new TGraph(tree1->GetSelectedRows(),tree1->GetV2(), tree1-
>GetV1());
gk->SetName(" ");
gk->Draw();
gk->SetLineColor(7);
gk->SetMarkerStyle(21);
gk->SetLineWidth(6);
gk->SetMarkerSize(7);
      mr->Add(gk);
      mr->Draw("A");
mr->GetXaxis()->SetTitle("Month of Year");
mr->GetYaxis()->SetTitle("Deposition Efficiency");
      TLegend *legend3=new TLegend(0.6,0.65,0.88,0.85);
      legend3->SetTextFont(50);
      legend3->SetTextSize(0.04);
      legend3->AddEntry(gi,"Deposition Efficiency 2001","l");
      legend3->AddEntry(gk,"Deposition Efficiency 2012","l");
      legend3->Draw();
r->Update();
r->Modified();

  }
{ TCanvas *e = new TCanvas("e", "Face 2",1400,800);
      TMultiGraph * mu = new TMultiGraph();
      tree2->Draw("fc:fa");
      TGraph *gu = new TGraph(tree2->GetSelectedRows(),tree2->GetV2(), tree2-
>GetV1());
gu->SetName(" ");
gu->Draw();
gu->SetLineColor(2);
gu->SetMarkerStyle(21);
gu->SetMarkerSize(7);
gu->SetLineWidth(6);
      mu->Add(gu);
      tree2->Draw("fm:fa");
      TGraph *gv = new TGraph(tree2->GetSelectedRows(),tree2->GetV2(), tree2-
>GetV1());
gv->SetName(" ");
gv->Draw();
gv->SetLineColor(3);
gv->SetMarkerStyle(21);
gv->SetLineWidth(6);
```

```
gv->SetMarkerSize(7);
        mu->Add(gv);
        mu->Draw("A");
mu->GetXaxis()->SetTitle("Month of Year");
mu->GetYaxis()->SetTitle("Aerosol Radius");
        TLegend *legend2=new TLegend(0.6,0.65,0.88,0.85);
        legend2->SetTextFont(50);
        legend2->SetTextSize(0.04);
        legend2->AddEntry(gu,"Aerosol raduis 2001","l");
        legend2->AddEntry(gv,"Aerosol radius 2002","l");
        legend2->Draw();
e->Update();
e->Modified();

    }

  }
```

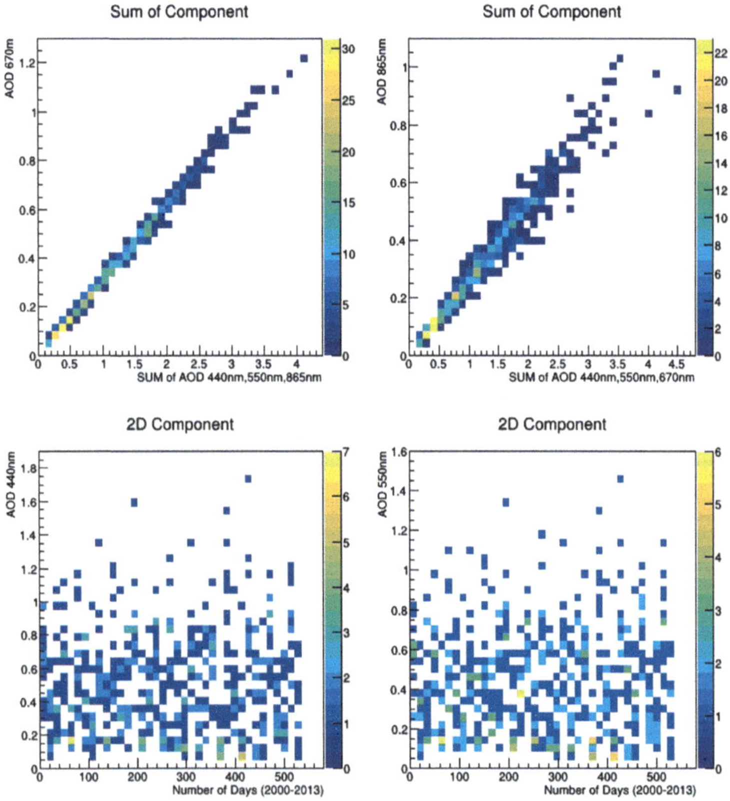

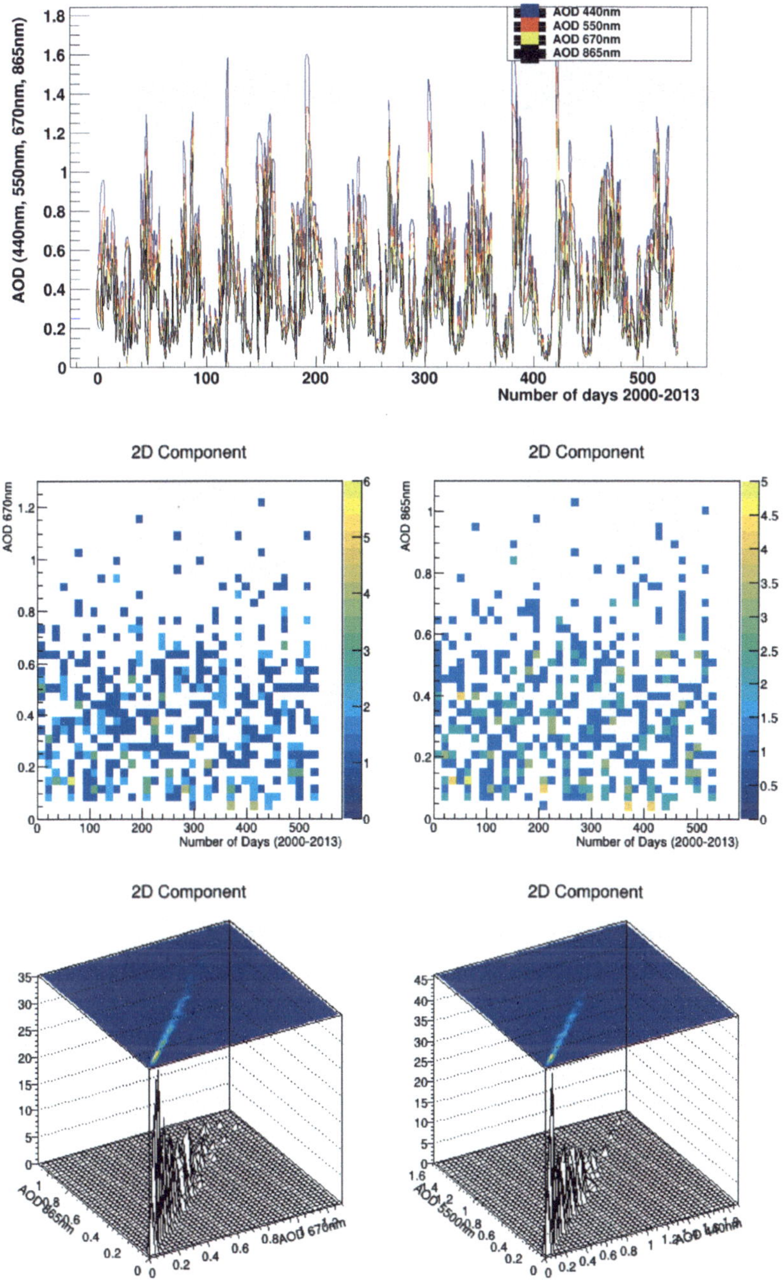

AOD 440nm
AOD 550nm
AOD 670nm
AOD 865nm
AOD (440nm, 550nm, 670nm, 865nm)
Number of days 2000-2013
2D Component
AOD 670nm
Number of Days (2000-2013)
2D Component
AOD 865nm
Number of Days (2000-2013)
2D Component
AOD 865nm
AOD 670nm
2D Component
AOD 440nm

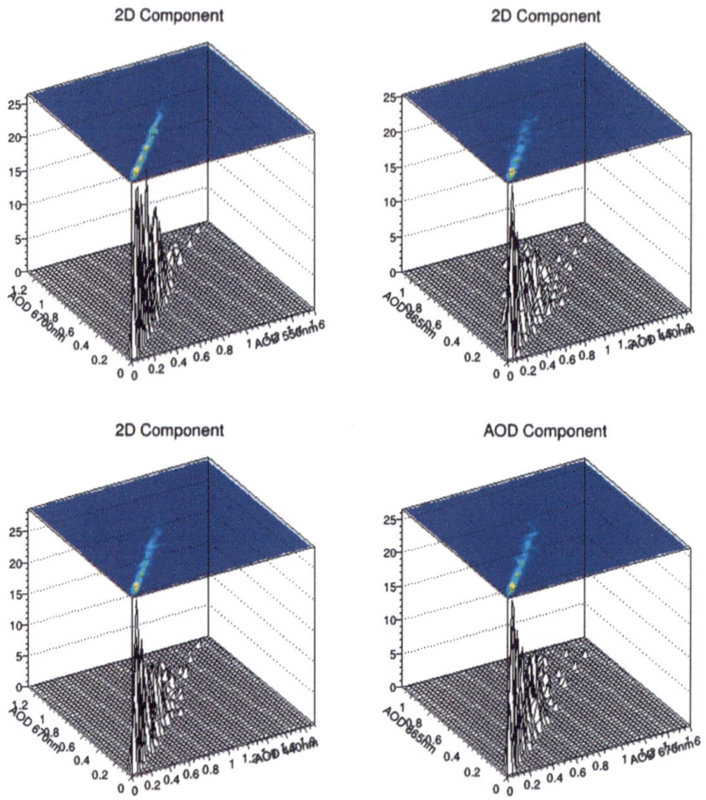

4.3 Designing C++ Codes to Analyze Big Data

So why does some modeller prefer the C++ code for analyzing big data? C++ is a compiled code that is capable of high performance. It has a flexible coding styles (functional, objectoriented, high level, low level) that it gives it the robustness to incorporate or link other like libraries for added functionality. C++ has powerful standard library with many functions that makes it efficient in analyzing any type of data with any type of complex equation. This process is achieved because C++ has local scoping of variables. Hence, many variables can be defined according to the users' comfort. These qualities give C++ a huge widespread adoption and support (i.e. crossplatform, industry, academia). When you have to process huge data collections rapidly and your calculation is not predefined, C++ can help. C++ is more prominent duplicate for reliability in solitary server. C++ will be the best dialect to utilize for dynamic load adjusting or an exceedingly proficient versatile storing layer. The strictness of using C++ is that pointers need to be used correctly and header files need to be complete before you can successfully run a code.

At the point when complex machine learning calculations are included, vast terabyte or petabyte of data can be prepared sensibly and rapidly using C++. This may be the reason why data scientist believe that C++ enhances processing speed. C++ is the main dialect where dataset above 1 GB+ are handled and meant to run within a duration of a second. Further, you can retrain and apply prescient investigation continuously to update or manipulate the RESTful API. Before it was revised in Java, Google's MapReduce was composed in C++. Further, MongoDB was likewise composed in C++. With regards to profound learning and profound neural systems, C++ is one of the program languages that can be utilized to compose profound learning calculations. Most of the deep learning algorithms require implementations in C++. For example, Caffe is a popular deep learning algorithm repository and machine learning software. Normally C and C++ applications require less limit and electric power than virtual machine dialects. This decreases CapEx and OpEx and server farm cost. C++ provides features or functionality that other languages need. Hence, it gives an extensive template to write generic code. In spite of the fact that the libraries accessible for C++ are not in the same class as the ones accessible for Java, a few data scientists have assembled their own (for instance Boost and Dlib).

The most vital things to note when using C++ to code for big data are to ensure that:

- the calculation that you are using for tackling the issue has sufficient functionality to unravel the technicality of the issue. Also, modeller can pay more attention about time limits and the most extreme loops in the program that need to be avoided;
- the data being utilized have the most extreme conceivable value i.e. given by the limitations of the issue. Words like 'int' are not used for 'big data' coding because it does not have the space to store such an expansive value. In this way, the modeller is advised to utilize 'long int';
- the 'long long int' and 'unsigned long int' are sufficient for taking care of the upper limits of larger dataset;
- the use of 'scanf' or 'printf' instead of 'cin' or 'cout' to avoid exceeding time limits. This is because 'cin' or 'cout' are slower than 'scanf' or 'printf' because they keep on maintaining data in a way, so that cin, cout can be interchangeably used with the c-style input/ouput i.e. scanf, printf. If you decide to use cin, cout only, then you can simply make it faster by switching off syncing with scanf, printf;
- an array is used to store larger volume of dataset.

Recently, some big data processing software are written in C++. For example, Thrill is an experimental Big Data processing framework written in C++. It is an university project from Karlsruhe Institute of Technology. Thrill utilizes much of the modern C++ features like lambdas and automatic type deduction. And of course, template meta-programming is not far away and in fact, Thrill relies heavily on it. The source code is available on Git Hub for further reading. a library of scalable algorithmic primitives like Map, Reduce By Key, Sort, and Window which can be combined efficiently to construct larger complex algorithms using pipelined data-flow style programming.

In designing C++ code, one of the added points that governs the execution of the codes i.e. source code, command and executable. In the command section, the compilers are used to convert the source codes into machine language or "code" that a computer's processor uses. Hence, the main reason for converting source code is to create an executable program. Bjarne [5] categorized the C++ compilers into free downloads and free downloads for trial period. The free downloads C++ compiler are: Apple C++ (for OS X devices); Bloodshed Dev-C++ (A GCC-based (Mingw) IDE); Clang C++ (with versions C++98/C++03, C++11, C++14 and C++17); Cygwin (GNU C++ for Unix-like environment and command-line interface for Microsoft Windows); Digital Mars C++ (for integrated development environment (IDE) for Windows and DOS); Mentor Graphics—Lite edition (for Siemens electronic design automation); MINGW Minimalist GNU for Windows); DJ Delorie's C++ development system for DOS/Windows (GNU C++); GNU CC source (for GNU Project supporting various programming languages); IBM C++ (for IBM power, System Z, Bluegene, and Cell); Intel C++ (for non-commercial development); Microsoft Visual C++ Express edition (set of integrated development environments developed by Microsoft as a freeware and registerware function-limited version of the non-free Microsoft Visual Studio); Oracle C++; Turbo C++; CINT (object oriented data analysis for package ROOT); LLVM; Open64; Comeau C/C++. The compiler can be expressed in Fig. 4.7.

The C++ compiler that works for trial period are Embarcadero C++, Edison Design Group C++ Front, Green Hills C++, HP C++, Intel C++; Mentor Graphics Sourcery CodeBench, Microsoft C++, Paradigm C++, The Portland Group C++, WindRiver's

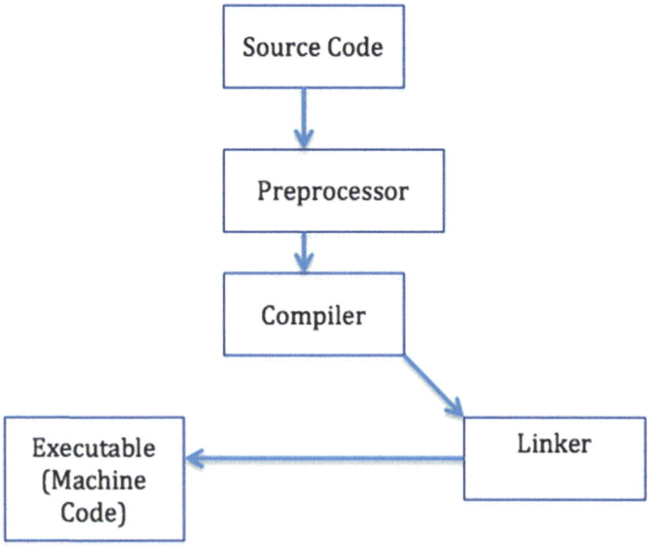

Fig. 4.7 The compiler unit of a computer

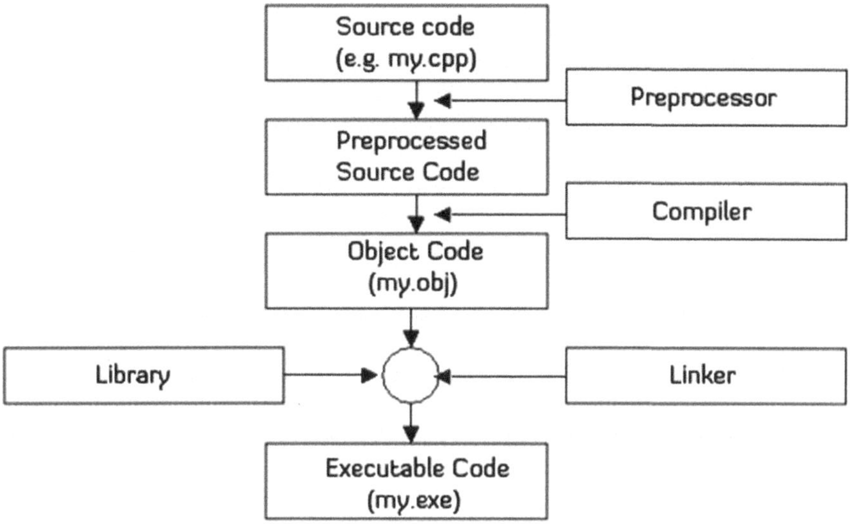

Fig. 4.8 C++ coding design [6]

Diab C++, EDGE C/C++, Microtec C/C++ etc. The illustration of the processes highlighted above can be further understood via Fig. 4.8.

C++ design pattern is an approach to portray and address by name (generally a shortsighted depiction of its objective), a repeatable arrangement or way to deal with a typical plan issue, that is, a typical method to take care of a non specific issue (how nonexclusive or complex, relies upon how confined the objective is). Pattern can develop individually or by plan. This is the reason configuration designs are valuable as a deliberation over the usage and an assistance at configuration organize. With this idea, a less demanding approach to encourage correspondence over a plan decision as standardization method is given so every individual can share the outline idea. The C++ design pattern issue depends on the kind of pattern they solve. The types of patterns are creational patterns, basic patterns, and standards of conduct.

Patterns are ordinarily found in protested arranged programming like C++ or Java. They can be viewed as a layout for how to take care of an issue that happens in a wide range of circumstances or applications. C++ big data design code can be effectively made from an outline design. Object-oriented design patterns regularly indicate connections and communications between classes or protests without determining the last application classes or questions that are involved.

Design patterns can speed up the development process by providing tested, proven development paradigms. Effective C++ design for big data analysis requires considering issues that may not become visible until later in the implementation. Reusing design patterns helps to prevent subtle issues that can cause major problems and improves code readability for coders and architects familiar with the patterns. Each outline design comprises of the accompanying parts: issue/necessity, forces and solution. Issue/necessity factor understudy a plan design that considers smaller than normal examination outline that might be coded to try out the design. This area expresses the prerequisites of the issue we need to understand.

Design patterns are regarded as an institutionalization of generally concurred prescribed procedures to take care of particular outline issues. One ought to comprehend them as an approach to execute great outline designs inside applications. Doing as such will lessen the utilization of wasteful and obscure solutions. Designs pattern accelerates your outline and imparts it to different developers. This section is focused on the reader to think broadly to start his/her C++ big data design codes. Hence, it is important to know the C++ inheritance. In heritance is the process by which a class can be derived from a base class with all features of base class and some of its own that makes it vital for code reusability. The types of inheritance with their pictorial interpretations are shown in Fig. 4.9.

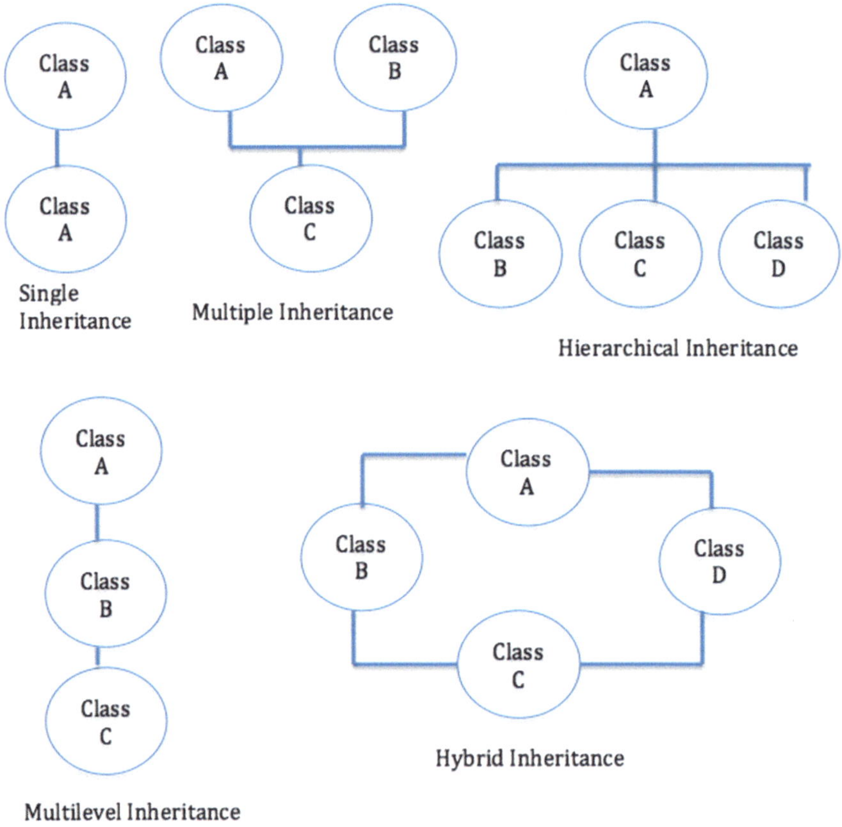

Fig. 4.9 Inheritance of C++ code design

References

1. Emetere ME (2016) Numerical modelling of West Africa regional scale aerosol dispersion. A doctoral thesis submitted to Covenant University, Nigeria, pp 65–289
2. Support (2017) Top real time big data applications in various domains. https://data-flair.training/blogs/big-data-applications-various-domains/. Accessed 29th Aug 2018
3. Cukier K (2010) Data, data everywhere. https://www.economist.com/special-report/2010/02/25/data-data-everywhere. Accessed 29th Aug 2018
4. Palmer WC (1965) Meteorological drought, Research Paper No. 45. US Weather Bureau, Washington, DC
5. Bjarne S (2018) An incomplete list of C++ compilers. http://www.stroustrup.com/compilers.html. Accessed 5th Nov 2018
6. SudoX (2018) C++ preprocessor directives. https://www.codingunit.com/cplusplus-tutorial-preprocessor-directives. Accessed 6th Sept 2018

Chapter 5
Data Treatment

Dataset processing began from nineteen-fifties in form of bookkeeping, balance sheet, cash flow statement, questionnaire, library documentation etc. Most of the manual data processing are supported by the application of mechanical or electronic calculators. Then development shifted the manual data processing into the automatic data processing were punched card equipment were adopted. This kind of data processing is referred to as mechanical data processing that is an advanced manual data processing. In this case, mechanical devices (typewriter, printing press, and weaving looms) were developed to help in automation of manual tasks. Later data processing technology evolved into the electronic data processing. The electronic data processing includes online processing, real time processing, distributed processing, time sharing, batch processing, multiprocessing, multitasking, interactive processing. Here, data processing was computerized with the use of electronic computers. Some field of study have redefined data processing as information technology. The commercial data processing is an off-shoot of the information technology. It entails the processing of large volume of input data, relatively few computational operations, and a large volume of output. Data processing alludes to the way toward changing crude information or data into significant outcome. Data processing is usually alluded to as cycle on the grounds that the yield acquired can be stored and utilized in future as information. Data is technically manipulated to produce results that lead to a resolution of a problem or improvement of an existing situation. Data Processing Cycle is a series of steps carried out to extract useful information from raw data (see Fig. 5.1). Although each step must be taken in order, the order is cyclic. The output and storage stage can lead to the repeat of the data collection stage, resulting in another cycle of data processing.

The cycle provides a view on how the data travels and transforms from collection to interpretation, and ultimately, used in effective business decisions.

The fundamental phases of information handling cycle are: Data collection; Data input; Data handling/ processing; Data yield/output. The procedure of data gathering or collection may include various stages relying upon the strategy utilized. These include:

M. E. Emetere and E. T. Akinlabi, *Introduction to Environmental Data Analysis and Modeling*, Lecture Notes in Networks and Systems 58,
https://doi.org/10.1007/978-3-030-36207-2_5

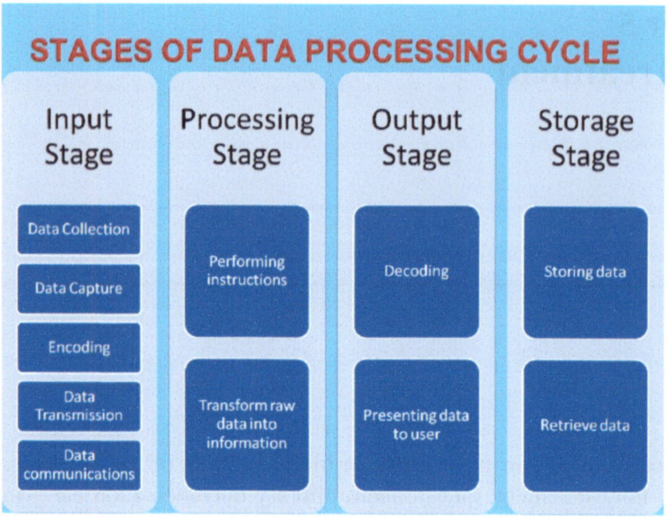

Fig. 5.1 Data processing cycle

i. Data creation: this is the procedure of recognizable proof and assembling certainties in a composed arrangement. This might be as physically arranged archive or caught from the source utilizing an information catch gadget. The success of data creation is directly proportional to the integrity of the filling system. Kelvin [1] highlighted the advantages of a good filling system as:

- it enables information to occupy much less space than the manual filing;
- it is much easier to update or modify information;
- it offers faster access and retrieval of data;
- it enhances data integrity and reduces duplication;
- it enhances security of data if proper care is taken to secure it.

These files can be organized on a storage media in the following format i.e.

- sequential,
- random,
- serial and
- indexed-sequential.

In data creation, data collection, data preparation, data input, data processing and data storage are very salient. Data Collection is the principal phase of the data processing cycle, and is exceptionally urgent, since the nature of information gathered will affect the outcome. The accumulation procedure needs to guarantee that the information assembled are both characterized and exact, with the goal that the discoveries are legitimate. Data preparation is the control of information into a frame appropriate to facilitate investigation and handling. Crude dataset cannot be handled and should be checked for exactness. Breaking down information that has not been

Fig. 5.2 Data processing procedure

painstakingly screened for issues can deliver exceedingly deceptive outcomes that are vigorously subject to the nature of information arranged. Data Input is where checked information is coded or changed over into machine language with the goal that of obtaining an expected outcome. Data can be inputted into the PC using a console, scanner, or information passage from a current source. This tedious procedure requires speed and precision. Data processing is the point at which the information is subjected to different means and strategies for great specialized controls utilizing Machine Learning and Artificial Intelligence calculations to create a yield or understanding about the information. The procedure might comprise of various strings of execution. Data storage is the last stage in the data processing cycle, where information and metadata (data about information) are saved or stored for future reference (Fig. 5.2).

ii. Data planning: this is the translation (transformation) of data from source report to machine decipherable frame.
iii. Data transmission: this will rely upon whether data should be transmitted by means of correspondence media to specific offices e.g. banking operation. In data processing or treatment, there are certain errors that must be avoided.

These errors include: Computational errors, Transcription errors; Transposition errors, Rounding errors, Overflow errors and Truncation errors [2].

a. Computational errors—Occurs when an arithmetic operation does not produce the expected results. The most common computation errors include overflow, truncation and rounding.
b. Overflow errors—Occurs if the result from a calculation is too large to be stored in the allocated memory space. For example if a byte is represented using 8 bits, an overflow will occur if the result of a calculation gives a 9-bit number.
c. Truncation errors—Result from having real numbers that have a long fractional part which cannot fit in the allocated memory space. The computer would truncate or cut off the extra characters from the fractional part. For example, a number like 0.784969 can be truncated to four digits to become 0.784.
d. Transcription errors—Occurs during data entry. Such errors include misreading and transposition errors. For example, a user may misread a handwritten figure such as 589 and type S89 instead i.e. confusing 5 for S.
e. Transposition errors—Results form incorrect arrangement of characters i.e. putting characters in the wrong order. For example, the user might enter 396 instead of 369.
f. Rounding errors—Results from raising or lowering a digit in a real number to the required rounded number. for example, to round off 30.666 to one decimal place we raise the first digit after the decimal point if its successor is more than or equal to five. In this case the successor is 6 therefore 30.666 rounded up to one decimal place is 30.7. If the successor is below five, e.g. 30.635, we round down the number to 30.6.

In environmental studies, data processing is very salient in communicating, interpreting and discussing results. It is possible to have a good data and communicate the result very poorly. Most good works are compromised because of poor data processing skills. In data processing, the following steps are very salient:

- Data integrity—Before data processing, there is the need to test the integrity of the data. Data integrity alludes to the precision and completeness of data entered in a PC or got from the data framework. Data integrity is estimated based on accuracy, timeliness and relevance of data. Accuracy of a data alludes to how close an estimation is to a real value. Computer is said to work on the principle of "garbage-in, garbage-out", hence, for whatever information is inputted into a computer, it deliver precise outcomes proficiently. Timeliness refers to relative precision of information in regard to the present situation for which it is required. This is critical in light of the fact that information and data have a time-value that relates both parameter. Relevance of a data refers to how valuable is the data to obtain an expected outcome. The value of the data implies that the information entered must be relevant to the preparing needs within reach and should meet the necessities of the handling cycle.
- Sorting of data—most satellite and ground dataset have lots of empty spaces or an inscription—"9999". The spaces or odd description are filtered out first. Then the data is reordered a predefined sequence. For example, the modeller may want to arrange the data in with respect to time, duration, number of occurrence, specific parameter etc. It should be noted here that in this regard that duration and time do

not connote same meaning. Time refers to the raw referral of time while duration refers to the predefined time related grouping of the modeller. There are numerous types of files used for storing data needed for processing, reference or back up. The main common types of processing files include Master files, transaction, reference, backup, report and sort file.

- Aggregation of data—There are time when the data is somewhat scattered such that there are pieces of information within same dataset or individual dataset. The modeller combines multiple pieces of data so that it will be easy to analyze the dataset in one piece. Also, scientist may which to import different parameters in various columns of different dataset into one dataset for comparative analysis. For example, if the scientist wants to compare satellite and ground dataset.
- Classification of data—This process ensures that the researcher separates dataset into various categories according to the research design. This process is almost similar to aggregation. The difference is that the aggregation does not follow a research plan while classification are in-line with the focus of the research.
- Validation of data—After sorting and aggregating dataset, it is important for the scientist or modeller to validate the dataset i.e. ensure that the supplied data is correct and relevant. Sometimes, dataset gotten from secondary sources are intentionally adulterated with wrong dataset. The validation process helps the researcher to understand the integrity of the dataset. What are the ways to validate dataset? In assessing the fundamentals of data validation, speculations can be made with respect to the diverse sorts of validation, as per the degree, intricacy, and motivation behind the different validation tasks to be done. For example, the type of validation includes: data type validation; range and constraint validation; code and cross-reference validation; and structured validation. The techniques for data validation includes: Source system loop back verification; Ongoing source-to-source verification; Data-Issue tracking; Data certification; Statistics collection; and Workflow management [3]. The Source system loop back verification entails that the researcher perform total based checks of theoretical background and guarantee that the data and theory are in agreement. For instance, theoretical knowledge of the dataset may have parabolic trend while the experimental dataset has a Gaussian trend. This may portend that the dataset may be faulty due to many reasons unknown. Ongoing source-to-source verification entails that the researcher presents the percentage error to guide the accuracy of the research. This is because an erroneous dataset would definitely translate to poor data analysis. Data-Issue tracking entails that the researcher diligently follows-up the sources of data error. This activity may help the researcher to discover repeating issues, uncover less secure branches of knowledge, and ensure the accuracy of the research. Data certification entails that the researcher performs in advance data validation before integrating the dataset (aggregation) to the main dataset. Statistics collection entails that the researcher keep up insights for the full life cycle of the dataset so that startling outcomes can be easily detected. Regardless of whether the researcher have an in-house-created measurements gathering procedure or depends upon metadata that has been imported from a secondary source, it is safe for the

researcher to set statistical alarms to detect such errors. Workflow management entails that the researcher considers data quality while planning dataset combination streams. The work processes in this technique can take into consideration issues rapidly and productively. Performing checks on data integrity gives the researcher further developed choices to determine the issue rapidly.

- Summarization of data—Sometimes, after doing the sorting, aggregation and validation of the data, the dataset may be still bulky to analyze. In this process, the dataset is averaged over minutes, hour, days, months or year.
- Analysis—This is the most important aspect of data treatment. Here, the dataset is organized, analyzed, interpreted and presented in different format such as plots, pie chart, bar chart, dummy plots etc. Data analysis uses specialized algorithms and statistical that are less regularly seen in an ordinary general business condition. There are data analytics tools that are used for analyzing data e.g. SPSS, SAS, DAP, gretl, PSPP etc.
- Reporting—This section is quite important as the data analysis. Here, the detail or summary of the analyzed dataset are discussed to reflect the overall message of the research.

5.1 Analogue Versus Digital Data Treatment Processes

This section is more of a hands-on section. The data treatment shall be done using the analogue and digital processes. The definition of the terms "analogue" and "digital" is not in its literal meaning. The definition is tied to the processing time to process a data.

5.1.1 The Analogue Data Treatment Processes

Find below the dataset used for the illustration. The details of the dataset are shown.

*********** Daily MIL3DAE.004 product ***********	
Selected Application: timeseries	
Aerosol Optical Depth at 443 nm (Blue Band) [unitless] (Optical_depth_average_Blue)	
Selected time period: 25 Feb 2000–30 Nov 2013	
Selected area for averaging: Lat (12.64, 12.65), Lon (−8, −7.99)	
Fill value: −9999.0	
Time (day)	*Optical_depth_average_Blue [unitless]*

(continued)

(continued)

25 Feb 2000	*−9.99900e+03*
26 Feb 2000	*−9.99900e+03*
27 Feb 2000	*−9.99900e+03*
28 Feb 2000	*−9.99900e+03*
29 Feb 2000	*−9.99900e+03*
01 Mar 2000	*−9.99900e+03*
02 Mar 2000	*−9.99900e+03*
03 Mar 2000	*−9.99900e+03*
04 Mar 2000	*−9.99900e+03*
05 Mar 2000	*−9.99900e+03*
06 Mar 2000	*−9.99900e+03*
07 Mar 2000	*−9.99900e+03*
08 Mar 2000	*0.775*
09 Mar 2000	*−9.99900e+03*
10 Mar 2000	*0.362*
11 Mar 2000	*−9.99900e+03*
12 Mar 2000	*−9.99900e+03*
13 Mar 2000	*−9.99900e+03*
14 Mar 2000	*−9.99900e+03*
15 Mar 2000	*−9.99900e+03*
16 Mar 2000	*−9.99900e+03*
17 Mar 2000	*0.487*
18 Mar 2000	*−9.99900e+03*
19 Mar 2000	*−9.99900e+03*
20 Mar 2000	*−9.99900e+03*
21 Mar 2000	*−9.99900e+03*
22 Mar 2000	*−9.99900e+03*
23 Mar 2000	*−9.99900e+03*
24 Mar 2000	*0.600*
25 Mar 2000	*−9.99900e+03*
26 Mar 2000	*0.520*
27 Mar 2000	*−9.99900e+03*
28 Mar 2000	*−9.99900e+03*
29 Mar 2000	*−9.99900e+03*
30 Mar 2000	*−9.99900e+03*
31 Mar 2000	*−9.99900e+03*
01 Apr 2000	*−9.99900e+03*
02 Apr 2000	*−9.99900e+03*

(continued)

(continued)

03 Apr 2000	−9.99900e+03
04 Apr 2000	−9.99900e+03
05 Apr 2000	−9.99900e+03
06 Apr 2000	−9.99900e+03
07 Apr 2000	−9.99900e+03
08 Apr 2000	−9.99900e+03
09 Apr 2000	0.557
10 Apr 2000	−9.99900e+03
11 Apr 2000	−9.99900e+03
12 Apr 2000	−9.99900e+03
13 Apr 2000	−9.99900e+03
14 Apr 2000	−9.99900e+03
15 Apr 2000	−9.99900e+03
16 Apr 2000	−9.99900e+03
17 Apr 2000	−9.99900e+03
18 Apr 2000	−9.99900e+03
19 Apr 2000	−9.99900e+03
20 Apr 2000	−9.99900e+03
21 Apr 2000	−9.99900e+03
22 Apr 2000	−9.99900e+03
23 Apr 2000	−9.99900e+03
24 Apr 2000	−9.99900e+03
25 Apr 2000	−9.99900e+03
26 Apr 2000	−9.99900e+03
27 Apr 2000	0.399
28 Apr 2000	−9.99900e+03
29 Apr 2000	−9.99900e+03
30 Apr 2000	−9.99900e+03
01 May 2000	−9.99900e+03
02 May 2000	−9.99900e+03
03 May 2000	−9.99900e+03
04 May 2000	0.403
05 May 2000	−9.99900e+03
06 May 2000	−9.99900e+03
07 May 2000	−9.99900e+03
08 May 2000	−9.99900e+03
09 May 2000	−9.99900e+03
10 May 2000	−9.99900e+03

(continued)

(continued)

11 May 2000	*−9.99900e+03*
12 May 2000	*−9.99900e+03*
13 May 2000	*0.551*
14 May 2000	*−9.99900e+03*
15 May 2000	*−9.99900e+03*
16 May 2000	*−9.99900e+03*
17 May 2000	*−9.99900e+03*
18 May 2000	*−9.99900e+03*
19 May 2000	*−9.99900e+03*
20 May 2000	*−9.99900e+03*
21 May 2000	*−9.99900e+03*
22 May 2000	*−9.99900e+03*
23 May 2000	*−9.99900e+03*
24 May 2000	*−9.99900e+03*
25 May 2000	*−9.99900e+03*
26 May 2000	*−9.99900e+03*
27 May 2000	*−9.99900e+03*
28 May 2000	*−9.99900e+03*
29 May 2000	*0.544*
30 May 2000	*−9.99900e+03*
31 May 2000	*−9.99900e+03*
01 Jun 2000	*−9.99900e+03*
02 Jun 2000	*−9.99900e+03*
03 Jun 2000	*−9.99900e+03*
04 Jun 2000	*−9.99900e+03*
05 Jun 2000	*0.455*
06 Jun 2000	*−9.99900e+03*
07 Jun 2000	*−9.99900e+03*
08 Jun 2000	*−9.99900e+03*
09 Jun 2000	*−9.99900e+03*
10 Jun 2000	*−9.99900e+03*
11 Jun 2000	*−9.99900e+03*
12 Jun 2000	*−9.99900e+03*
13 Jun 2000	*−9.99900e+03*
14 Jun 2000	*0.457*
15 Jun 2000	*−9.99900e+03*
16 Jun 2000	*−9.99900e+03*
17 Jun 2000	*−9.99900e+03*

(continued)

(continued)

18 Jun 2000	*−9.99900e+03*
19 Jun 2000	*−9.99900e+03*
20 Jun 2000	*−9.99900e+03*
21 Jun 2000	*0.648*
22 Jun 2000	*−9.99900e+03*
23 Jun 2000	*−9.99900e+03*
24 Jun 2000	*−9.99900e+03*
25 Jun 2000	*−9.99900e+03*
26 Jun 2000	*−9.99900e+03*
27 Jun 2000	*−9.99900e+03*
28 Jun 2000	*0.296*
29 Jun 2000	*−9.99900e+03*
30 Jun 2000	*−9.99900e+03*
01 Jul 2000	*−9.99900e+03*
02 Jul 2000	*−9.99900e+03*
03 Jul 2000	*−9.99900e+03*
04 Jul 2000	*−9.99900e+03*
05 Jul 2000	*−9.99900e+03*
06 Jul 2000	*−9.99900e+03*
07 Jul 2000	*0.283*
08 Jul 2000	*−9.99900e+03*
09 Jul 2000	*−9.99900e+03*
10 Jul 2000	*−9.99900e+03*
11 Jul 2000	*−9.99900e+03*
12 Jul 2000	*−9.99900e+03*
13 Jul 2000	*−9.99900e+03*
14 Jul 2000	*0.674*
15 Jul 2000	*−9.99900e+03*
16 Jul 2000	*−9.99900e+03*
17 Jul 2000	*−9.99900e+03*
18 Jul 2000	*−9.99900e+03*
19 Jul 2000	*−9.99900e+03*
20 Jul 2000	*−9.99900e+03*
21 Jul 2000	*−9.99900e+03*
22 Jul 2000	*−9.99900e+03*
23 Jul 2000	*0.263*
24 Jul 2000	*−9.99900e+03*
25 Jul 2000	*−9.99900e+03*

(continued)

(continued)

26 Jul 2000	*−9.99900e+03*
27 Jul 2000	*−9.99900e+03*
28 Jul 2000	*−9.99900e+03*
29 Jul 2000	*−9.99900e+03*
30 Jul 2000	*−9.99900e+03*
31 Jul 2000	*−9.99900e+03*
01 Aug 2000	*−9.99900e+03*
02 Aug 2000	*−9.99900e+03*
03 Aug 2000	*−9.99900e+03*
04 Aug 2000	*−9.99900e+03*
05 Aug 2000	*−9.99900e+03*
06 Aug 2000	*−9.99900e+03*
07 Aug 2000	*−9.99900e+03*
08 Aug 2000	*−9.99900e+03*
09 Aug 2000	*−9.99900e+03*
10 Aug 2000	*−9.99900e+03*
11 Aug 2000	*−9.99900e+03*
12 Aug 2000	*−9.99900e+03*
13 Aug 2000	*−9.99900e+03*
14 Aug 2000	*−9.99900e+03*
15 Aug 2000	*−9.99900e+03*
16 Aug 2000	*−9.99900e+03*
17 Aug 2000	*0.402*
18 Aug 2000	*−9.99900e+03*
19 Aug 2000	*−9.99900e+03*
20 Aug 2000	*−9.99900e+03*
21 Aug 2000	*−9.99900e+03*
22 Aug 2000	*−9.99900e+03*
23 Aug 2000	*−9.99900e+03*
24 Aug 2000	*0.170*
25 Aug 2000	*−9.99900e+03*
26 Aug 2000	*−9.99900e+03*
27 Aug 2000	*−9.99900e+03*
28 Aug 2000	*−9.99900e+03*
29 Aug 2000	*−9.99900e+03*
30 Aug 2000	*−9.99900e+03*
31 Aug 2000	*−9.99900e+03*
01 Sep 2000	*−9.99900e+03*

(continued)

(continued)

02 Sep 2000	*0.337*
03 Sep 2000	*−9.99900e+03*
04 Sep 2000	*−9.99900e+03*
05 Sep 2000	*−9.99900e+03*
06 Sep 2000	*−9.99900e+03*
07 Sep 2000	*−9.99900e+03*
08 Sep 2000	*−9.99900e+03*
09 Sep 2000	*−9.99900e+03*
10 Sep 2000	*−9.99900e+03*
11 Sep 2000	*−9.99900e+03*
12 Sep 2000	*−9.99900e+03*
13 Sep 2000	*−9.99900e+03*
14 Sep 2000	*−9.99900e+03*
15 Sep 2000	*−9.99900e+03*
16 Sep 2000	*0.358*
17 Sep 2000	*−9.99900e+03*
18 Sep 2000	*−9.99900e+03*
19 Sep 2000	*−9.99900e+03*
20 Sep 2000	*−9.99900e+03*
21 Sep 2000	*−9.99900e+03*
22 Sep 2000	*−9.99900e+03*
23 Sep 2000	*−9.99900e+03*
24 Sep 2000	*−9.99900e+03*
25 Sep 2000	*0.271*
26 Sep 2000	*−9.99900e+03*
27 Sep 2000	*−9.99900e+03*
28 Sep 2000	*−9.99900e+03*
29 Sep 2000	*−9.99900e+03*
30 Sep 2000	*−9.99900e+03*
01 Oct 2000	*−9.99900e+03*
02 Oct 2000	*−9.99900e+03*
03 Oct 2000	*−9.99900e+03*
04 Oct 2000	*0.442*
05 Oct 2000	*−9.99900e+03*
06 Oct 2000	*−9.99900e+03*

(continued)

(continued)

07 Oct 2000	*−9.99900e+03*
08 Oct 2000	*−9.99900e+03*
09 Oct 2000	*−9.99900e+03*
10 Oct 2000	*−9.99900e+03*
11 Oct 2000	*−9.99900e+03*
12 Oct 2000	*−9.99900e+03*
13 Oct 2000	*−9.99900e+03*
14 Oct 2000	*−9.99900e+03*
15 Oct 2000	*−9.99900e+03*
16 Oct 2000	*−9.99900e+03*
17 Oct 2000	*−9.99900e+03*
18 Oct 2000	*0.694*
19 Oct 2000	*−9.99900e+03*
20 Oct 2000	*0.359*
21 Oct 2000	*−9.99900e+03*
22 Oct 2000	*−9.99900e+03*
23 Oct 2000	*−9.99900e+03*
24 Oct 2000	*−9.99900e+03*
25 Oct 2000	*−9.99900e+03*
26 Oct 2000	*−9.99900e+03*
27 Oct 2000	*−9.99900e+03*
28 Oct 2000	*−9.99900e+03*
29 Oct 2000	*−9.99900e+03*
30 Oct 2000	*−9.99900e+03*
31 Oct 2000	*−9.99900e+03*
01 Nov 2000	*−9.99900e+03*
02 Nov 2000	*−9.99900e+03*
03 Nov 2000	*−9.99900e+03*
04 Nov 2000	*−9.99900e+03*
05 Nov 2000	*0.301*
06 Nov 2000	*−9.99900e+03*
07 Nov 2000	*−9.99900e+03*
08 Nov 2000	*−9.99900e+03*
09 Nov 2000	*−9.99900e+03*
10 Nov 2000	*−9.99900e+03*
11 Nov 2000	*−9.99900e+03*
12 Nov 2000	*−9.99900e+03*
13 Nov 2000	*−9.99900e+03*

(continued)

(continued)

14 Nov 2000	*−9.99900e+03*
15 Nov 2000	*−9.99900e+03*
16 Nov 2000	*−9.99900e+03*
17 Nov 2000	*−9.99900e+03*
18 Nov 2000	*−9.99900e+03*
19 Nov 2000	*−9.99900e+03*
20 Nov 2000	*−9.99900e+03*
21 Nov 2000	*0.138*
22 Nov 2000	*−9.99900e+03*
23 Nov 2000	*−9.99900e+03*
24 Nov 2000	*−9.99900e+03*
25 Nov 2000	*−9.99900e+03*
26 Nov 2000	*−9.99900e+03*
27 Nov 2000	*−9.99900e+03*
28 Nov 2000	*0.214*
29 Nov 2000	*−9.99900e+03*
30 Nov 2000	*−9.99900e+03*
01 Dec 2000	*−9.99900e+03*
02 Dec 2000	*−9.99900e+03*
03 Dec 2000	*−9.99900e+03*
04 Dec 2000	*−9.99900e+03*
05 Dec 2000	*−9.99900e+03*
06 Dec 2000	*−9.99900e+03*
07 Dec 2000	*−9.99900e+03*
08 Dec 2000	*−9.99900e+03*
09 Dec 2000	*−9.99900e+03*
10 Dec 2000	*−9.99900e+03*
11 Dec 2000	*−9.99900e+03*
12 Dec 2000	*−9.99900e+03*
13 Dec 2000	*−9.99900e+03*
14 Dec 2000	*−9.99900e+03*
15 Dec 2000	*−9.99900e+03*
16 Dec 2000	*−9.99900e+03*
17 Dec 2000	*−9.99900e+03*
18 Dec 2000	*−9.99900e+03*
19 Dec 2000	*−9.99900e+03*
20 Dec 2000	*−9.99900e+03*
21 Dec 2000	*0.133*

(continued)

(continued)

22 Dec 2000	*−9.99900e+03*
23 Dec 2000	*−9.99900e+03*
24 Dec 2000	*−9.99900e+03*
25 Dec 2000	*−9.99900e+03*
26 Dec 2000	*−9.99900e+03*
27 Dec 2000	*−9.99900e+03*
28 Dec 2000	*−9.99900e+03*
29 Dec 2000	*−9.99900e+03*
30 Dec 2000	*0.134*
31 Dec 2000	*−9.99900e+03*

Step 1: Fit the data into columns.

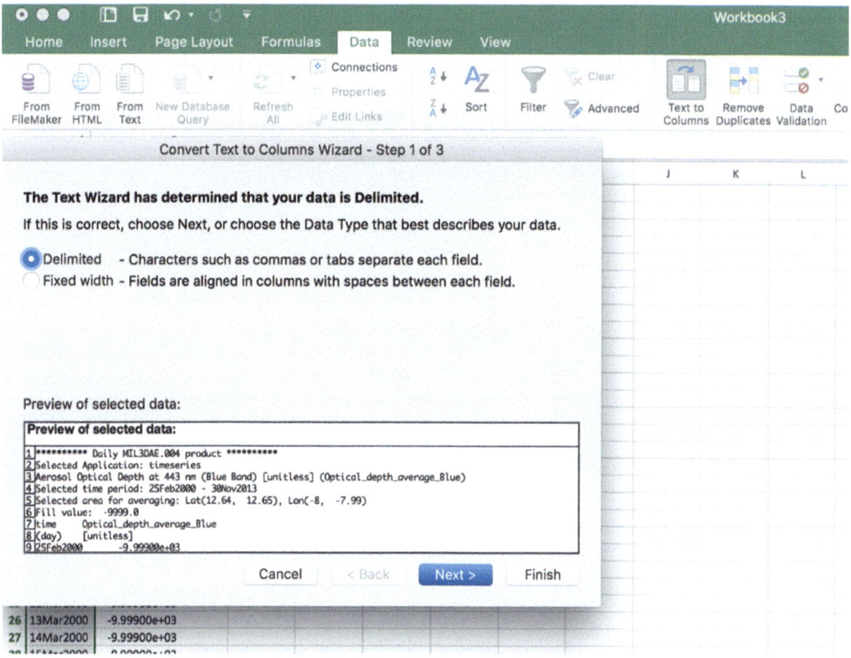

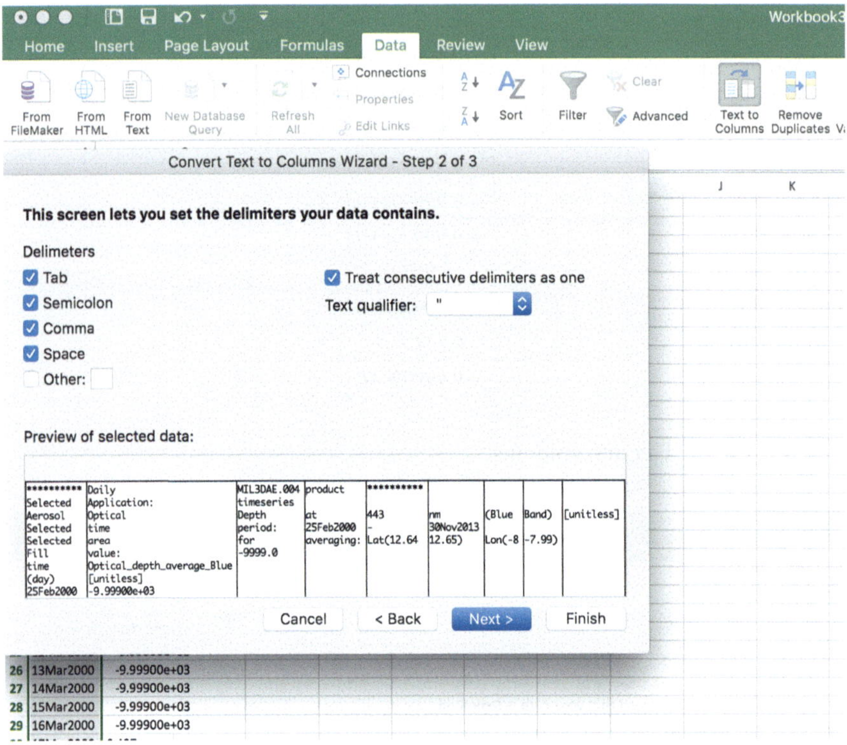
Workbook3
Home Insert Page Layout Formulas Data Review View
From FileMaker
From HTML
From Text
New Database Query
Refresh All
Connections
Properties
Edit Links
Sort
Filter
Clear
Advanced
Text to Columns
Remove Duplicates
Convert Text to Columns Wizard - Step 2 of 3
This screen lets you set the delimiters your data contains.
Delimeters
Tab
Semicolon
Comma
Space
Other:
Treat consecutive delimiters as one
Text qualifier: "
Preview of selected data:
Cancel
< Back
Next >
Finish
26 13Mar2000 -9.99900e+03
27 14Mar2000 -9.99900e+03
28 15Mar2000 -9.99900e+03
29 16Mar2000 -9.99900e+03

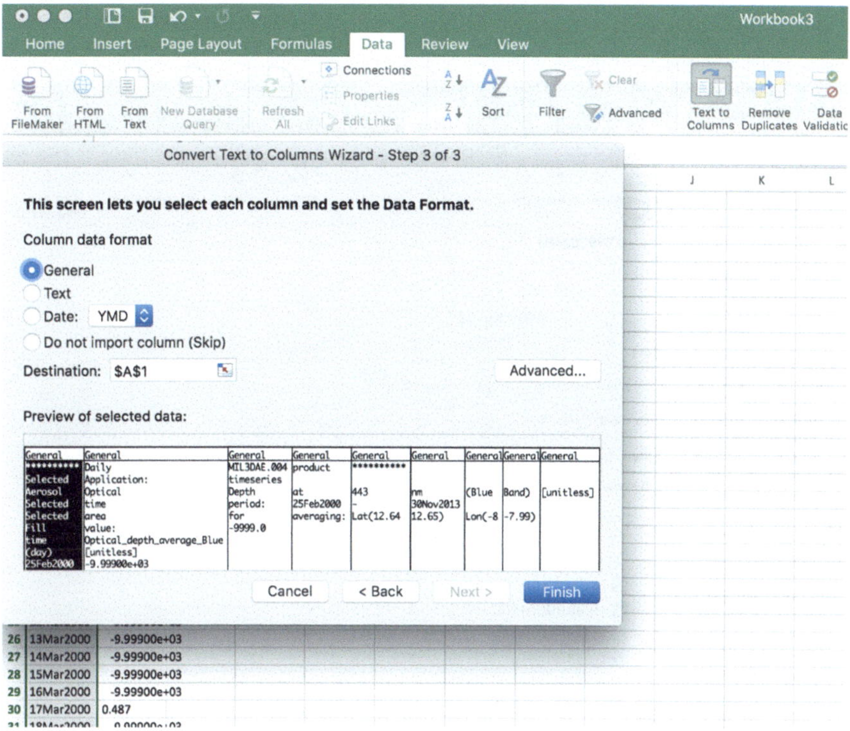

Step 2: Highlight the columns.

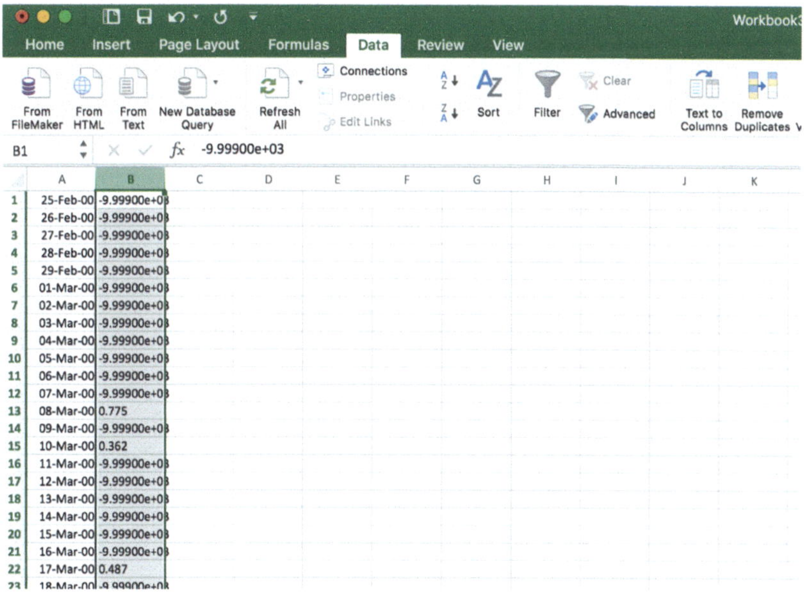

Step 3: Filter the data to remove the missing data "*−9.99900e+03*".

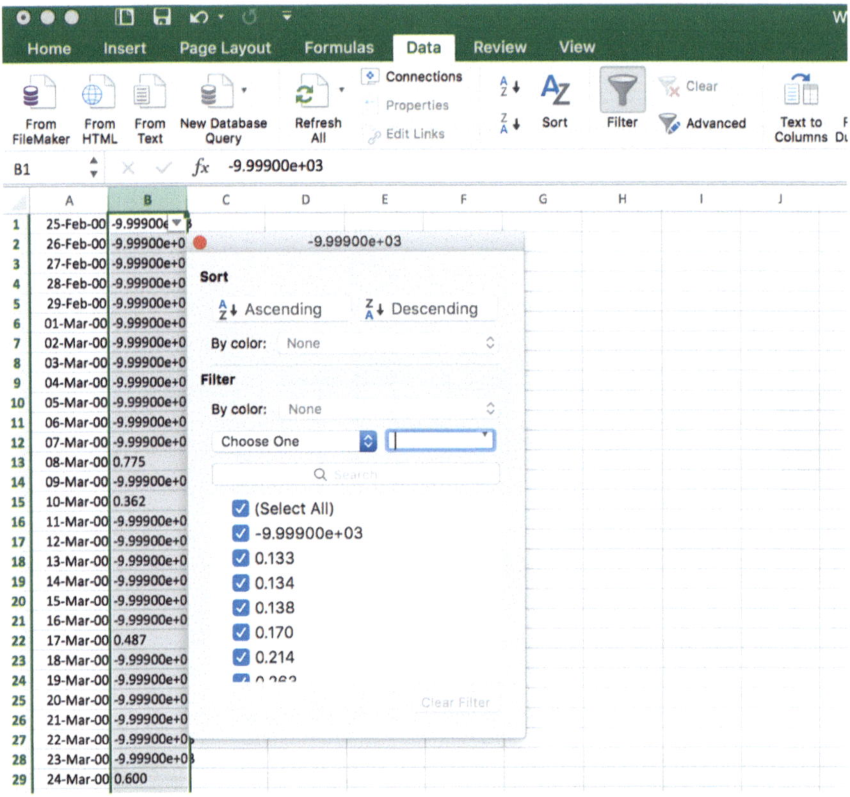

Step 4: Copy out the filtered data

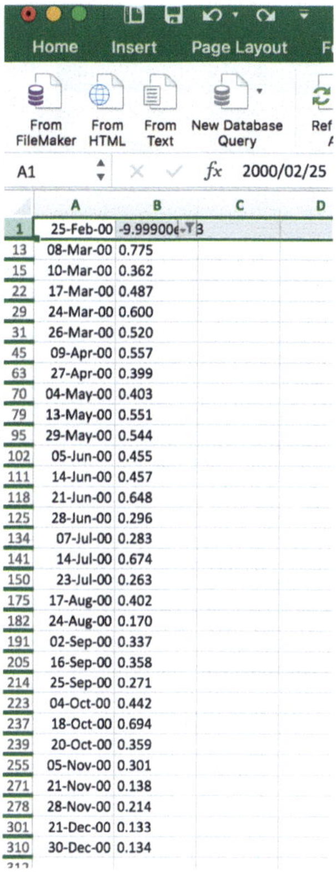

Step 4: Find the average for each month

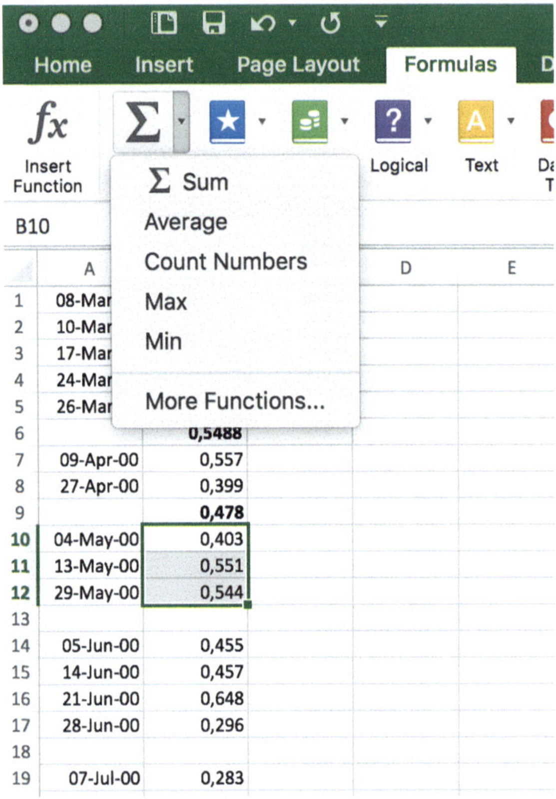

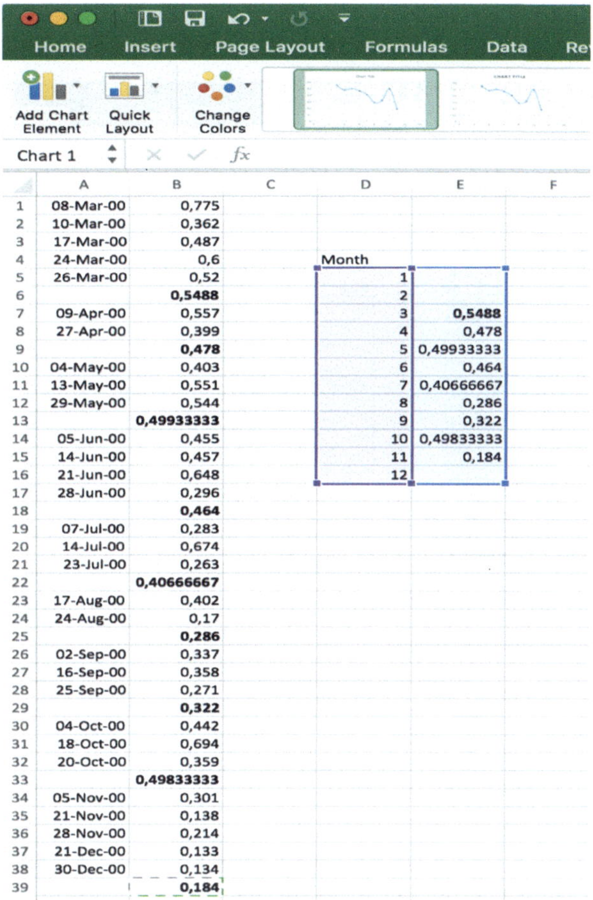
Home
Insert
Page Layout
Formulas
Data
Add Chart Element
Quick Layout
Change Colors
Chart 1
A B C D E F
1 08-Mar-00 0,775
2 10-Mar-00 0,362
3 17-Mar-00 0,487
4 24-Mar-00 0,6 Month
5 26-Mar-00 0,52 1
6 0,5488 2
7 09-Apr-00 0,557 3 0,5488
8 27-Apr-00 0,399 4 0,478
9 0,478 5 0,49933333
10 04-May-00 0,403 6 0,464
11 13-May-00 0,551 7 0,40666667
12 29-May-00 0,544 8 0,286
13 0,49933333 9 0,322
14 05-Jun-00 0,455 10 0,49833333
15 14-Jun-00 0,457 11 0,184
16 21-Jun-00 0,648 12
17 28-Jun-00 0,296
18 0,464
19 07-Jul-00 0,283
20 14-Jul-00 0,674
21 23-Jul-00 0,263
22 0,40666667
23 17-Aug-00 0,402
24 24-Aug-00 0,17
25 0,286
26 02-Sep-00 0,337
27 16-Sep-00 0,358
28 25-Sep-00 0,271
29 0,322
30 04-Oct-00 0,442
31 18-Oct-00 0,694
32 20-Oct-00 0,359
33 0,49833333
34 05-Nov-00 0,301
35 21-Nov-00 0,138
36 28-Nov-00 0,214
37 21-Dec-00 0,133
38 30-Dec-00 0,134
39 0,184

Step 5: Plot the graph

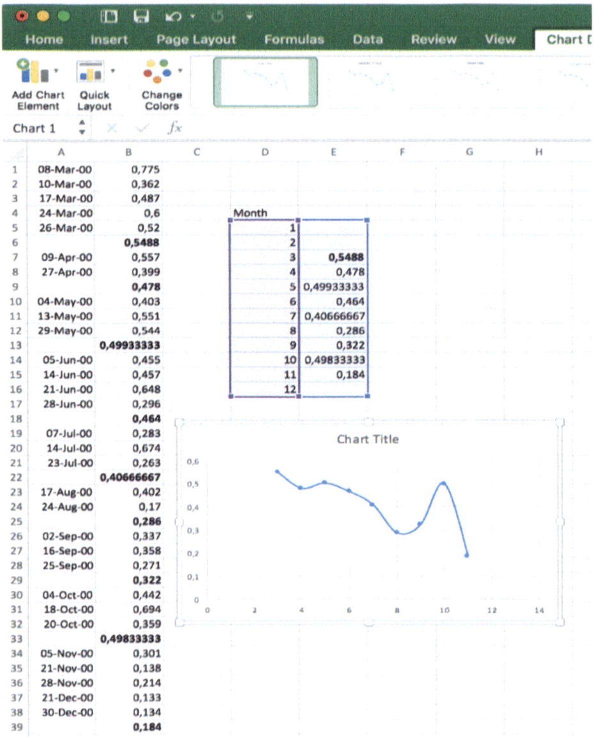

From the above exercise, it is clearly shown that it will be quite difficult to process large pieces of the data using this method. This method takes time, concentration (due to fatigue), prone to mistakes (by the user), over working of the computer and prolong storage time. In this age of huge data, it will be extremely difficult for modern researcher to adopt this method. These, therefore, necessitate need to switch to the digital data treatment process.

5.1.2 *The Digital Data Treatment Processes*

From the previous chapters, the need to automate data assimilation, treatment and analysis was systematically emphasized. In this section, the focus is to illustrate flow chart for writing your own script with any preferred language. The general rule in planning the script writing is

i. Write the header files. Most programming language will certainly require that you write the header file. This process helps your script to locate the libraries attached to your programme.

ii. Link libraries. This part may be optional because some libraries are sophisticated and process your commands adequately. However, there are times that specific libraries are needed to aid your work. For example, in most programming language, there are many emerging libraries that makes the programmer use less lines of command.
iii. Indicate the type of data streaming you want. This aspect is important because if the wrong data streaming command is used, the loop command line to obtain new data may be redundant. For example, if you want your script to automatically locate the data location in your computer (without you typing the data address every time you run the script), there must be a stream command that enables the programmer to link the data addresses that has been defined in a folder or file.
iv. Write the main body of the command line. The writing of the command line is at the discretion of the researcher and the programming language that has been adopted.
v. Create a location to store the treated dataset. When writing the script, it is expected that the researcher specifies the location where treated data will be stored. A neatly planned script must provide the specific naming of each file that has been processed so that the raw data and the processed data will have the same filename.

The chart describing the script design for treating dataset is shown in Fig. 5.3.

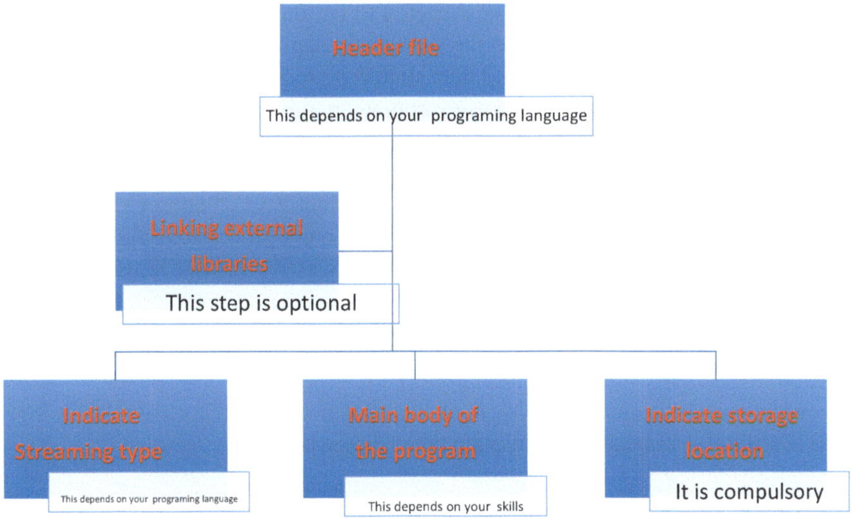

Fig. 5.3 Planning script writing to treat dataset

5.2 Designing a Data Treatment Scheme for Large Work Flow

In previous section, the need to plan a script writing was emphasized. In this section, the focus is assist the reader plan for data treatment scheme for large work flow. In addition to the point listed in previous section, the below points are important when working on huge data.

i. The statistical data treatment. This step is basic to understand the end goal utilizing the information in the correct frame. Statistical data treatment is basic in all investigations, regardless of the discipline involved. Measurable treatment of data enormously relies upon the sort of analysis and the coveted outcome from the statistical-test. It shows the need to treat data after an experiment; the collation of data is similarly essential so accurate inferences can be drawn from the data. There are numerous methods engaged with measurements that treat data in the required way. It provides an intensive comprehension of data to enhance direct or correct tests with the correct inductions from the information got.
ii. Determination of errors in the data. All experiments invariably produce errors and noise. Both systematic and random errors need to be taken into consideration.
iii. Listing of all file location. In order to examine all the required datasets in whatsoever folders, the researcher is expected to list out the file and its respective location. The pitfall to avoid is the systematic naming of the output files. The process of listing out the treated data should preferably be by the original filename of the raw dataset. If you use automatic naming (e.g. 'dataset 1', 'dataset 2' etc.), you may eventual have problem when re-using the treated file. This is because the scripts see the numbering differently. For example, the script sees 'dataset 11', 'dataset 12', 'dataset 13', 'dataset 14', 'dataset 15', 'dataset 16', 'dataset 17', 'dataset 18', 'dataset 19', before it processes 'dataset 2'. This arrangement, your original 'dataset 2' will be logically 'dataset 11' in the output file.

The new chart becomes for designing data treatment for large dataset is shown in Fig. 5.4.

For readers who prefer the analogue of data treatment, it is important to show how to do statistical data treatment using the table shown below.

Month	AOD 2005	AOD 2006	AOD 2007
Jan	0.268	0.101	0.291
Feb	0.337	0.161	0.148
Mar	0.163	0.202	0.230

(continued)

(continued)

Month	AOD 2005	AOD 2006	AOD 2007
Apr	0.171	0.338	0.120
May	0.733	0.270	0.464
Jun	0.497	0.501	0.511
Jul	0.584	0.686	0.655
Aug	0.811	0.681	0.419
Sep	0.503	0.744	0.396
Oct	0.356	0.256	0.381
Nov	0.167	0.274	0.162
Dec	0.160	0.145	0.146

The outlook of the dataset is given below

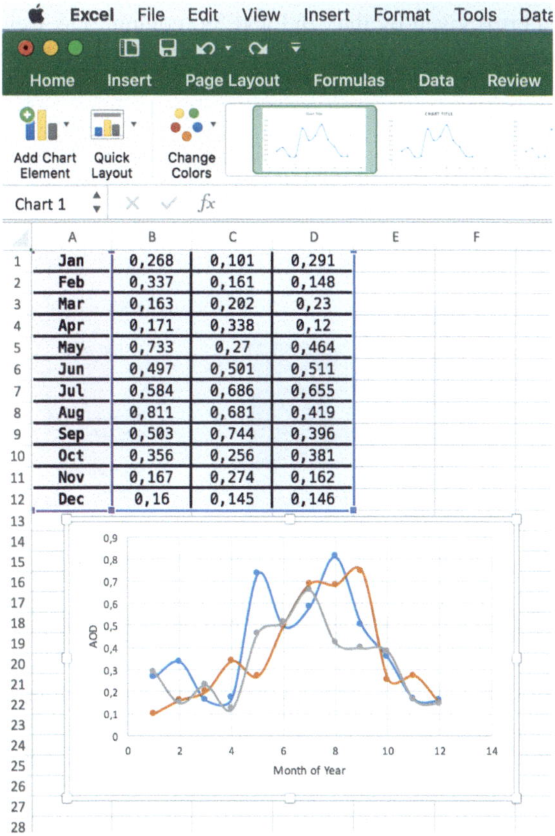

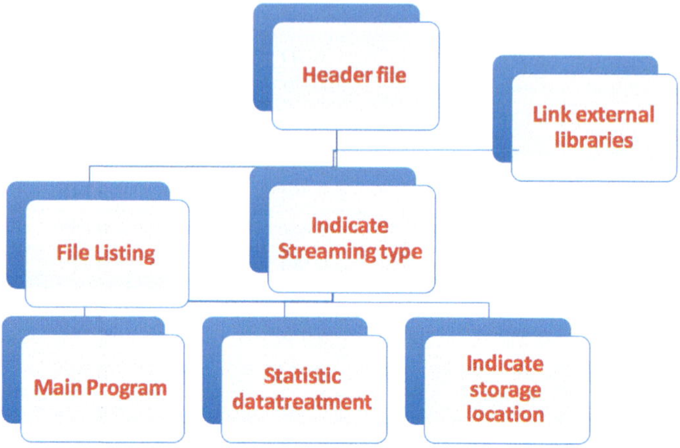

Fig. 5.4 Data treatment design for large dataset

The combined error given by the '2nd and 3rd column' has weighted mean of 0.30757 and standard error of 0.06708. The diagram below show the combined error in pictorial form

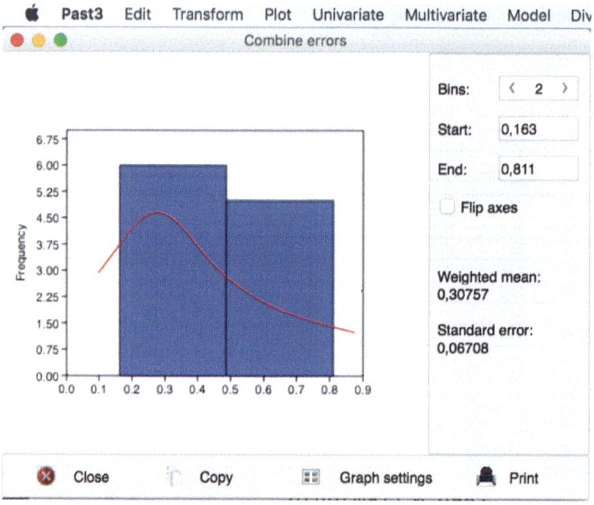

Likewise, the combined error given by the '3rd and 4th column' has weighted mean of 0.2875 and standard error of 0.068409. The diagram below show the combined error in pictorial form

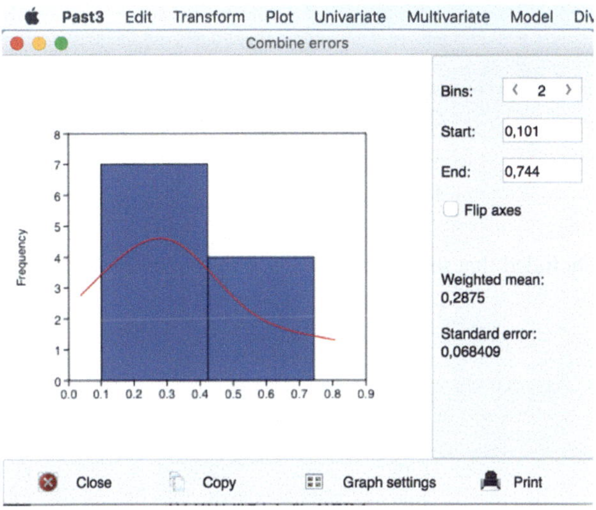

The T-Test, Wilcoxon test and Normalcy test are presented below

T-test			
Parameter	AOD 2005	AOD 2006	AOD 2007
Given mean	0.396	0.363	0.327
Sample mean	0.39583	0.36325	0.34336
95% conf. interval	(0.25028–0.54139)	(0.2172–0.5093)	(0.22971–0.45702)
Difference	0.00016667	0.00025	0.016364
95% conf. interval	(−0.14539 to 0.14572)	(−0.1458 to 0.1463)	(−0.097294 to 0.13002)
t	−0.0025202	0.0037676	0.32079
p (same mean)	0.99803	0.99706	0.75497

It can be concluded that the means are not significantly different

Wilcoxon test			
Parameter	AOD 2005	AOD 2006	AOD 2007
Given median	0.347	0.272	0.336
Sample median	0.3465	0.272	0.381
W	43	47.5	33
Normal appr. z	0.31379	0.66705	0

(continued)

(continued)

Wilcoxon test			
Parameter	AOD 2005	AOD 2006	AOD 2007
p (same median)	0.75368	0.50474	1
p (exact)	0.79102	0.53076	1

It can be concluded that the medians are not significantly different

Normality test			
Parameter	AOD 2005	AOD 2006	AOD 2007
N	12	12	11
Shapiro-Wilk W	0.8936	0.8651	0.9542
p(normal)	0.131	0.05669	0.6982
Anderson-Darling A	0.4693	0.6993	0.2298
p (normal)	0.2017	0.04979	0.7456
p (Monte Carlo)	0.2145	0.0507	0.7752
Jarque-Bera JB	1.06	1.442	0.4751
p (normal)	0.5887	0.4863	0.7886
p (Monte Carlo)	0.2721	0.1485	0.7112

5.3 Application of Numerical Methods in Image Analysis

Sizeable environmental dataset appears as image form. This image has specific information that may be deeper than its facial interpretation. Image analysis involves processing an image into fundamental components in order to extract statistical data. When analyzing an image, the emphasis may be finding shapes, detecting edges, removing noise, counting objects, and measuring region and image properties of an object. Hence, image analysis may be image enhancement to remove noise; image segmentation to isolate regions and objects of interest; morphological filtering to remove more noise; region analysis to extract statistical data. In this section, the image below is used to generate outcomes.

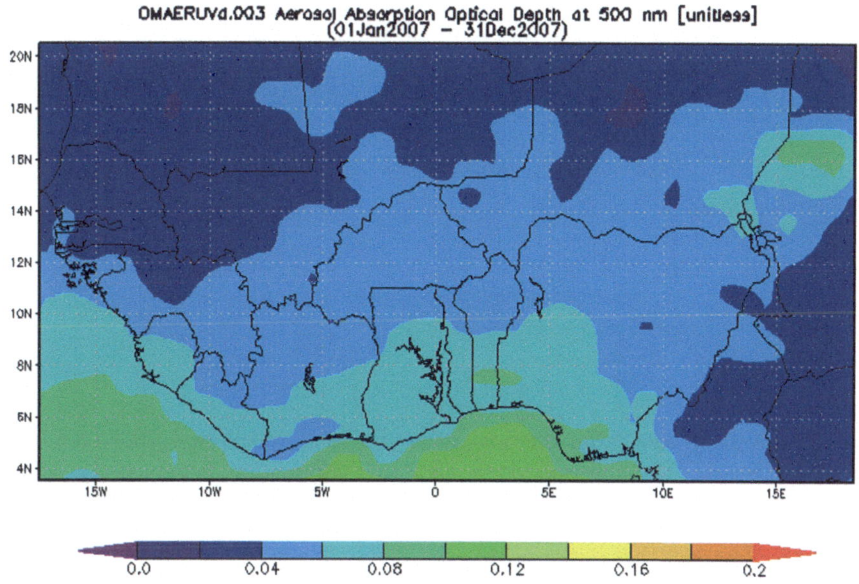

In this exercise, the MultiSpecWin64_2018.08.30 software was used to perform the morphological filtering as shown below.

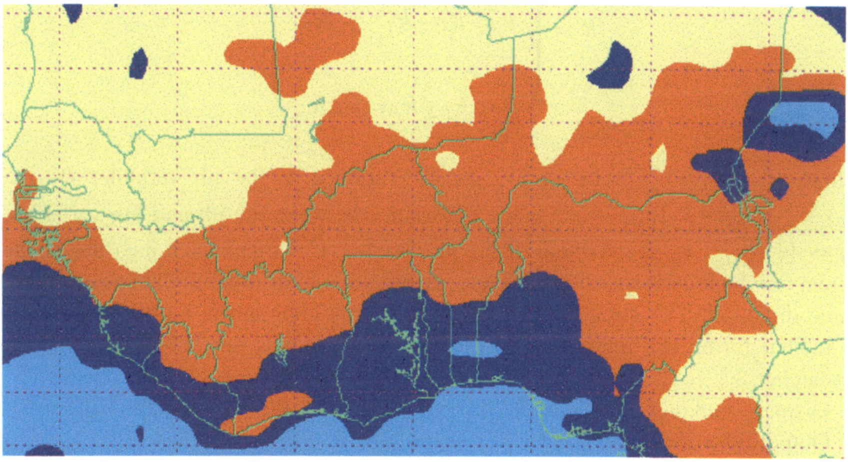

Image histogram is an operation that computes the histogram for an image or any part of an image as defined by an ROI. The graph below shows the histogram of the raw image above.

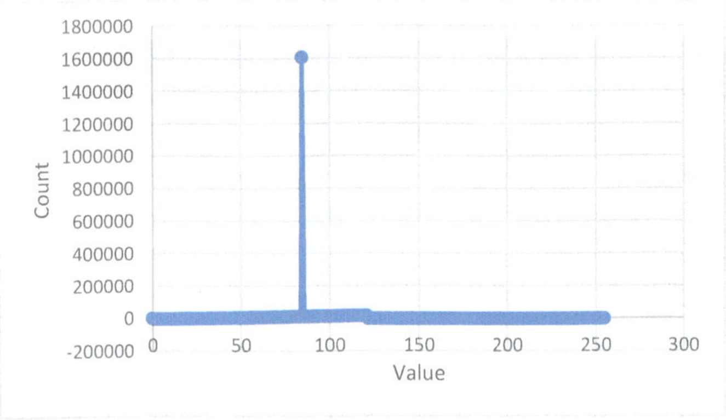

The statistical data from the image on the MultiSpecWin64_2018.08.30 is presented below. The information is the raw information extracted by MultiSpecWin64_2018.08.30. One of the analysis that is done via the MultiSpecWin64_2018.08.30 software is covariance matrix. Covariance matrix is referred to as high-level statistics. The covariance matrix refers to the symmetric array of numbers. It is mathematically represented as:

$$R = \begin{pmatrix} R_1^2 & R_{12} & R_{13} & \dots & R_{1P} \\ R_{21} & R_2^2 & R_{23} & \dots & R_{2P} \\ R_{31} & R_{32} & R_3^2 & \dots & R_{3P} \\ \vdots & \vdots & \vdots & \ddots & \vdots \\ R_{P1} & R_{P2} & R_{P3} & \dots & R_P^2 \end{pmatrix}$$

where $R_j^2 = \left(\frac{1}{n}\right)\sum_{i=1}^{n}\left(x_{ij} - \bar{x}_j\right)^2$ is the variance of the jth variable, $R_{jk} = \left(\frac{1}{n}\right)\sum_{i=1}^{n}\left(x_{ij} - \bar{x}_j\right)(x_{ik} - \bar{x}_k)$ is the covariance between the jth and kth variables, $\bar{x}_j = \left(\frac{1}{n}\right)\sum_{i=1}^{n} x_{ij}$ is the mean of the jth variable. Covariance is a measure of the extent to which corresponding elements from two sets of ordered data move in the same direction. The numerical analysis of the raw diagram is highlighted below in tabular form. The interpretation of the tables depends on the expertise of the researcher.

Display 'sat8.png' 09-15-2018 23:09:01 (MultiSpecWin64_2018.08.30)

Output Information:

Statistics computed from base image file and saved in 'sat8.sta'

Number of Bad Lines = 0.

Total Number of Pixels = 785,920.

Histogram Summary Table

First line	Last line	Line interval	First column	Last column	Column interval
1	640	1	1	1228	1

Description	Data range	Channel mean	Channel median	Standard deviation
1	0–170	32.6	33	17.3
2	0–218	135.1	162	58.9
3	0–252	223.6	251	56.8
4	255–255	255.0	255	0.0

Principal Components 09-15-2018 23:09:51 (MultiSpecWin64_2018.08.30)
Input Parameters:
Image file = ‘sat8.png’.
Correlation matrix will be used.
For selected area.

Lines: 1 to 640 by 1
Columns: 1 to 1228 by 1

Channels used: 1–4.
Output Information:
785,920 pixels used for covariance matrix.
0 CPU seconds for computing covariance matrix.
After 8 iterations, the largest off diagonal was 1.0e−13.

Component	Eigenvalue	%	Cum. %	‘Cum.’ ‘determinant’	‘Cum.’ ‘Log Det.’	Feature ‘mean’
1	1.1396	37.9855	37.9855	1.1396E+00	1.3065E−01	88.2211
2	1.0225	34.0841	72.0696	1.1652E+00	1.5292E−01	114.5594
3	0.8379	27.9304	100.0000	9.7636E−01	−2.3923E−02	220.0547
4	0.0000	0.0000	100.0000	9.7636E−01	−2.3923E−02	255.0000

Eigenvectors component	Channel			
	1	2	3	4
1	0.28180	−0.62286	0.72982	0.00000
2	0.86501	0.49404	0.08764	0.00000
3	−0.41515	0.60660	0.67800	0.00000
4	0.00000	0.00000	0.00000	1.00000

Description information for: 'sat8.png'

File format	PNG
Image type	Multispectral
Band interleave format	BIS
Data type	Integer
Signed data	No
Compression algorithm	Lossless-deflate
Number of lines	640
Number of columns	1228
Number of channels	4
Number of bytes	1
Number of bits	8
Number of header bytes	0
Number of pre-line bytes	0
Number of post-line bytes	0
Number of pre-channel bytes	0
Number of post-channel bytes	0
Block height	1
Block width	1228
Line start	1
Column start	1

Planar Map Parameters

x-map coordinate for center of upper-left pixel (1)	1.000000
y-map coordinate for center of upper-left pixel (1)	1.000000
horizontal pixel size per pixel	0.0000000000
vertical pixel size per pixel	0.0000000000

MultiSpec Classification
Cluster 09-15-2018 23:12:18 (MultiSpecWin64_2018.08.30)
Input Parameters:
Project = 'Untitled Project'.
Base image file = 'sat8.png'.
Image start line and column = 1, 1.
Cluster algorithm is "ISODATA".
Initialize within eigenvector volume.
Number Clusters: 10.

Convergence (percent): 99.0.
Minimum cluster size: 5.
Clusters from selected area.

Lines: 1 to 640 by 1
Columns: 1 to 1228 by 1

Channels used: 1–4.
Output Information:

Initial cluster	Cluster pixels	Class %	Channel means			
			1	2	3	4
1	0	0.0	32.9	108.9	249.9	255.0
2	0	0.0	40.9	104.6	228.1	255.0
3	0	0.0	31.3	124.4	245.1	255.0
4	0	0.0	38.4	102.1	194.9	255.0
5	0	0.0	46.7	141.2	241.3	255.0
6	0	0.0	56.4	114.0	173.1	255.0
7	0	0.0	29.5	165.7	220.9	255.0
8	0	0.0	35.0	159.3	175.5	255.0
9	0	0.0	35.0	196.1	198.4	255.0
10	0	0.0	29.6	210.1	185.6	255.0

Final cluster	Cluster pixels	Class %	Channel means			
			1	2	3	4
1	269,272	34.3	35.0	70.0	251.0	255.0
2	5164	0.7	109.0	30.0	216.0	255.0
3	292,116	37.2	27.0	162.0	252.0	255.0
4	22,420	2.9	0.0	0.0	0.0	255.0
5	8988	1.1	170.0	170.0	170.0	255.0
6	119,164	15.2	33.0	200.0	199.0	255.0
7	68,796	8.8	33.2	211.0	119.1	255.0

Cluster	Channel standard deviations			
	1	2	3	4
1	0.0	0.0	0.0	0.0
2	0.0	0.0	0.0	0.0
3	0.0	0.0	0.0	0.0
4	0.0	0.0	0.0	0.0
5	0.0	0.0	0.0	0.0
6	0.0	0.0	0.0	0.0
7	0.4	3.7	42.8	0.0

Feature Extraction 09-15-2018 23:13:01 (MultiSpecWin64_2018.08.30)
Input Parameters:
Project = 'Untitled Project'.
Original class statistics are used.
Base image file = 'sat8.png'.
Channels used: 1–4.

Classes used	Weight
1: Cluster1	10.000
2: Cluster2	10.000
3: Cluster3	10.000
4: Cluster4	10.000
5: Cluster5	10.000
6: Cluster6	10.000
7: Cluster7	10.000

Check Covariances 09-15-2018 23:13:45 (MultiSpecWin64_2018.08.30)
Input Parameters:
Original class statistics are used.
Image file = 'sat8.png'.
Channels used: 1–4.
Output Information:
Class 1: Cluster1
At least one variance value in covariance matrix for class 'Cluster1' was zero and set to 0.010.
Original Covariance Matrix

	Channel			
	1	2	3	4
1	1.0000e–02	0.0000e+00	0.0000e+00	0.0000e+00
2	0.0000e+00	1.0000e–02	0.0000e+00	0.0000e+00
3	0.0000e+00	0.0000e+00	1.0000e–02	0.0000e+00
4	0.0000e+00	0.0000e+00	0.0000e+00	1.0000e–02

Inverted Covariance Matrix.
Det = 1.000000000e−08.
Log Det = −18.4206807.
Rank of matrix = 4.
Smallest pivot value = 0.01.
0 CPU seconds for matrix inversion.

	Channel			
	1	2	3	4
1	1.0000e+02	0.0000e+00	0.0000e+00	0.0000e+00
2	0.0000e+00	1.0000e+02	0.0000e+00	0.0000e+00
3	0.0000e+00	0.0000e+00	1.0000e+02	0.0000e+00
4	–0.0000e+00	–0.0000e+00	–0.0000e+00	1.0000e+02

Multiplied matrices.
Total error = 0.000000000e+00.
Average error = 0.000000000e+00.
Maximum error = 0.00000e+00.

	Channel			
	1	2	3	4
1	1.0000e+00	0.0000e+00	0.0000e+00	0.0000e+00
2	0.0000e+00	1.0000e+00	0.0000e+00	0.0000e+00
3	0.0000e+00	0.0000e+00	1.0000e+00	0.0000e+00
4	0.0000e+00	0.0000e+00	0.0000e+00	1.0000e+00

Inverted-Inverted Covariance Matrix.
Det = 1.000000000e+08.
Log Det = 18.4206807.

	Channel			
	1	2	3	4
1	1.0000e–02	0.0000e+00	0.0000e+00	0.0000e+00
2	0.0000e+00	1.0000e–02	0.0000e+00	0.0000e+00
3	0.0000e+00	0.0000e+00	1.0000e–02	0.0000e+00
4	–0.000e+00	–0.000e+00	–0.000e+00	1.0000e–02

Class 2: Cluster2

At least one variance value in covariance matrix for class 'Cluster2' was zero and set to 0.010.

Original Covariance Matrix

	Channel			
	1	2	3	4
1	1.0000e–02	0.0000e+00	0.0000e+00	0.0000e+00
2	0.0000e+00	1.0000e–02	0.0000e+00	0.0000e+00
3	0.0000e+00	0.0000e+00	1.0000e–02	0.0000e+00
4	0.0000e+00	0.0000e+00	0.0000e+00	1.0000e–02

Inverted Covariance Matrix.

Det = 1.000000000e–08.

Log Det = –18.4206807.

Rank of matrix = 4.

Smallest pivot value = 0.01.

	Channel			
	1	2	3	4
1	1.0000e+02	0.0000e+00	0.0000e+00	0.0000e+00
2	0.0000e+00	1.0000e+02	0.0000e+00	0.0000e+00
3	0.0000e+00	0.0000e+00	1.0000e+02	0.0000e+00
4	-0.000e+00	-0.000e+00	-0.000e+00	1.0000e+02

Multiplied matrices.

Total error = 0.000000000e+00.

Average error = 0.000000000e+00.

Maximum error = 0.00000e+00.

	Channel			
	1	2	3	4
1	1.0000e+00	0.0000e+00	0.0000e+00	0.0000e+00
2	0.0000e+00	1.0000e+00	0.0000e+00	0.0000e+00
3	0.0000e+00	0.0000e+00	1.0000e+00	0.0000e+00
4	0.0000e+00	0.0000e+00	0.0000e+00	1.0000e+00

Inverted-Inverted Covariance Matrix.

Det = 1.000000000e+08.

Log Det = 18.4206807.

	Channel			
	1	2	3	4
1	1.0000e–02	0.0000e+00	0.0000e+00	0.0000e+00
2	0.0000e+00	1.0000e–02	0.0000e+00	0.0000e+00
3	0.0000e+00	0.0000e+00	1.0000e–02	0.0000e+00
4	–0.000e+00	–0.000e+00	–0.000e+00	1.0000e–02

Class 3: Cluster3

At least one variance value in covariance matrix for class 'Cluster3' was zero and set to 0.010.

Original Covariance Matrix

	Channel			
	1	2	3	4
1	1.0000e–02	0.0000e+00	0.0000e+00	0.0000e+00
2	0.0000e+00	1.0000e–02	0.0000e+00	0.0000e+00
3	0.0000e+00	0.0000e+00	1.0000e–02	0.0000e+00
4	0.0000e+00	0.0000e+00	0.0000e+00	1.0000e–02

Inverted Covariance Matrix.

Det = 1.000000000e−08

Log Det = −18.4206807.

Rank of matrix = 4.

Smallest pivot value = 0.01.

	Channel			
	1	2	3	4
1	1.0000e+02	0.0000e+00	0.0000e+00	0.0000e+00
2	0.0000e+00	1.0000e+02	0.0000e+00	0.0000e+00
3	0.0000e+00	0.0000e+00	1.0000e+02	0.0000e+00
4	–0.0000e+00	–0.0000e+00	–0.0000e+00	1.0000e+02

Multiplied matrices.

Total error = 0.000000000e+00.

Average error = 0.000000000e+00.

Maximum error = 0.00000e+00.

	Channel			
	1	2	3	4
1	1.0000e+00	0.0000e+00	0.0000e+00	0.0000e+00
2	0.0000e+00	1.0000e+00	0.0000e+00	0.0000e+00
3	0.0000e+00	0.0000e+00	1.0000e+00	0.0000e+00
4	0.0000e+00	0.0000e+00	0.0000e+00	1.0000e+00

Inverted-Inverted Covariance Matrix.

Det = 1.000000000e+08.

Log Det = 18.4206807.

	Channel			
	1	2	3	4
1	1.0000e–02	0.0000e+00	0.0000e+00	0.0000e+00
2	0.0000e+00	1.0000e–02	0.0000e+00	0.0000e+00
3	0.0000e+00	0.0000e+00	1.0000e–02	0.0000e+00
4	–0.000e+00	–0.000e+00	–0.000e+00	1.0000e–02

Class 4: Cluster4

At least one variance value in covariance matrix for class 'Cluster4' was zero and set to 0.010.

Original Covariance Matrix

	Channel			
	1	2	3	4
1	1.0000e-02	0.0000e+00	0.0000e+00	0.0000e+00
2	0.0000e+00	1.0000e-02	0.0000e+00	0.0000e+00
3	0.0000e+00	0.0000e+00	1.0000e-02	0.0000e+00
4	0.0000e+00	0.0000e+00	0.0000e+00	1.0000e-02

Inverted Covariance Matrix.

Det = 1.000000000e−08.

Log Det = −18.4206807.

Rank of matrix = 4.

Smallest pivot value = 0.01.

	Channel			
	1	2	3	4
1	1.0000e+02	0.0000e+00	0.0000e+00	0.0000e+00
2	0.0000e+00	1.0000e+02	0.0000e+00	0.0000e+00
3	0.0000e+00	0.0000e+00	1.0000e+02	0.0000e+00
4	–0.000e+00	–0.000e+00	–0.000e+00	1.0000e+02

Multiplied matrices.

Total error = 0.000000000e+00.

Average error = 0.000000000e+00.

Maximum error = 0.00000e+00.

	Channel			
	1	2	3	4
1	1.0000e+00	0.0000e+00	0.0000e+00	0.0000e+00
2	0.0000e+00	1.0000e+00	0.0000e+00	0.0000e+00
3	0.0000e+00	0.0000e+00	1.0000e+00	0.0000e+00
4	0.0000e+00	0.0000e+00	0.0000e+00	1.0000e+00

Inverted-Inverted Covariance Matrix.

Det = 1.000000000e+08.

Log Det = 18.4206807.

	Channel			
	1	2	3	4
1	1.0000e–02	0.0000e+00	0.0000e+00	0.0000e+00
2	0.0000e+00	1.0000e–02	0.0000e+00	0.0000e+00
3	0.0000e+00	0.0000e+00	1.0000e–02	0.0000e+00
4	–0.000e+00	–0.000e+00	–0.000e+00	1.0000e–02

Class 5: Cluster5

At least one variance value in covariance matrix for class 'Cluster5' was zero and set to 0.010.

Original Covariance Matrix

	Channel			
	1	2	3	4
1	1.0000e–02	0.0000e+00	0.0000e+00	0.0000e+00
2	0.0000e+00	1.0000e–02	0.0000e+00	0.0000e+00
3	0.0000e+00	0.0000e+00	1.0000e–02	0.0000e+00
4	0.0000e+00	0.0000e+00	0.0000e+00	1.0000e–02

Inverted Covariance Matrix.

Det = 1.000000000e−08.

Log Det = −18.4206807.

Rank of matrix = 4.

Smallest pivot value = 0.01.

0 CPU seconds for matrix inversion.

	Channel			
	1	2	3	4
1	1.0000e+02	0.0000e+00	0.0000e+00	0.0000e+00
2	0.0000e+00	1.0000e+02	0.0000e+00	0.0000e+00
3	0.0000e+00	0.0000e+00	1.0000e+02	0.0000e+00
4	–0.0000e+00	–0.0000e+00	–0.0000e+00	1.0000e+02

Multiplied matrices.

Total error = 0.000000000e+00.

Average error = 0.000000000e+00.

Maximum error = 0.00000e+00.

	Channel			
	1	2	3	4
1	1.0000e+00	0.0000e+00	0.0000e+00	0.0000e+00
2	0.0000e+00	1.0000e+00	0.0000e+00	0.0000e+00
3	0.0000e+00	0.0000e+00	1.0000e+00	0.0000e+00
4	0.0000e+00	0.0000e+00	0.0000e+00	1.0000e+00

Inverted-Inverted Covariance Matrix.

Det = 1.000000000e+08.

Log Det = 18.4206807.
0 CPU seconds for matrix inversion.

	Channel			
	1	2	3	4
1	1.0000e–02	0.0000e+00	0.0000e+00	0.0000e+00
2	0.0000e+00	1.0000e–02	0.0000e+00	0.0000e+00
3	0.0000e+00	0.0000e+00	1.0000e–02	0.0000e+00
4	–0.000e+00	–0.000e+00	–0.000e+00	1.0000e–02

Class 6: Cluster6
At least one variance value in covariance matrix for class 'Cluster6' was zero and set to 0.010.
Original Covariance Matrix

	Channel			
	1	2	3	4
1	1.0000e–02	0.0000e+00	0.0000e+00	0.0000e+00
2	0.0000e+00	1.0000e–02	0.0000e+00	0.0000e+00
3	0.0000e+00	0.0000e+00	1.0000e–02	0.0000e+00
4	0.0000e+00	0.0000e+00	0.0000e+00	1.0000e–02

Inverted Covariance Matrix.
Det = 1.000000000e–08.
Log Det = –18.4206807.
Rank of matrix = 4.
Smallest pivot value = 0.01.
0 CPU seconds for matrix inversion.

	Channel			
	1	2	3	4
1	1.0000e+02	0.0000e+00	0.0000e+00	0.0000e+00
2	0.0000e+00	1.0000e+02	0.0000e+00	0.0000e+00
3	0.0000e+00	0.0000e+00	1.0000e+02	0.0000e+00
4	–0.0000e+00	–0.0000e+00	–0.0000e+00	1.0000e+02

Multiplied matrices.
Total error = 0.000000000e+00.
Average error = 0.000000000e+00.
Maximum error = 0.00000e+00.

	Channel			
	1	2	3	4
1	1.0000e+00	0.0000e+00	0.0000e+00	0.0000e+00
2	0.0000e+00	1.0000e+00	0.0000e+00	0.0000e+00
3	0.0000e+00	0.0000e+00	1.0000e+00	0.0000e+00
4	0.0000e+00	0.0000e+00	0.0000e+00	1.0000e+00

Inverted-Inverted Covariance Matrix.
Det = 1.000000000e+08.
Log Det = 18.4206807.
0 CPU seconds for matrix inversion.

	Channel			
	1	2	3	4
1	1.0000e–02	0.0000e+00	0.0000e+00	0.0000e+00
2	0.0000e+00	1.0000e–02	0.0000e+00	0.0000e+00
3	0.0000e+00	0.0000e+00	1.0000e–02	0.0000e+00
4	–0.000e+00	–0.000e+00	–0.000e+00	1.0000e–02

Class 7: Cluster7

At least one variance value in covariance matrix for class 'Cluster7' was zero and set to 0.010.

Original Covariance Matrix

	Channel			
	1	2	3	4
1	1.7300e−01	1.5570e+00	−1.7819e+01	0.0000e+00
2	1.5570e+00	1.4013e+01	−1.6037e+02	0.0000e+00
3	−1.7819e+01	−1.6037e+02	1.8354e+03	0.0000e+00
4	0.0000e+00	0.0000e+00	0.0000e+00	1.0000e−02

Inverted Covariance Matrix.
Det = 8.881094988e−11.
Log Det = −23.1445112.
Rank of matrix = 2.
Smallest pivot value = 4.83879603e−12.

	Channel			
	1	2	3	4
1	8.0047e−14	5.2047e−13	9.7087e−03	0.0000e+00
2	5.2047e−13	4.8388e−12	8.7379e−02	0.0000e+00

(continued)

(continued)

	Channel			
	1	2	3	4
3	−9.7087e−03	−8.7379e−02	5.4484e−04	0.0000e+00
4	0.0000e+00	0.0000e+00	−0.0000e+00	1.0000e+02

Multiplied matrices.
Total error = 2.106143894e+02.
Average error = 1.316339934e+01.
Maximum error = 1.60374e+02.

	Channel			
	1	2	3	4
1	1.7300e−01	1.5570e+00	1.2802e−01	0.0000e+00
2	1.5570e+00	1.4013e+01	1.1522e+00	0.0000e+00
3	−1.7819e+01	−1.6037e+02	−1.3186e+01	0.0000e+00
4	0.0000e+00	0.0000e+00	0.0000e+00	1.0000e+00

Inverted-Inverted Covariance Matrix.
Det = −1.696258366e−14.
Log Det = −31.7077664.

	Channel			
	1	2	3	4
1	2.2217e−14	1.9652e−13	−1.1111e−01	0.0000e+00
2	−1.9652e−13	1.1444e+01	−7.1361e−02	0.0000e+00
3	1.1111e−01	−7.1361e−02	1.1444e+01	0.0000e+00
4	−0.0000e+00	0.0000e+00	0.0000e+00	1.0000e−02

Aside the use of software to analyze environmental image dataset, the researcher can write scripts that can analyze image. This method is advantageous when working on huge dataset. Scripts can be used to remove noise as below.

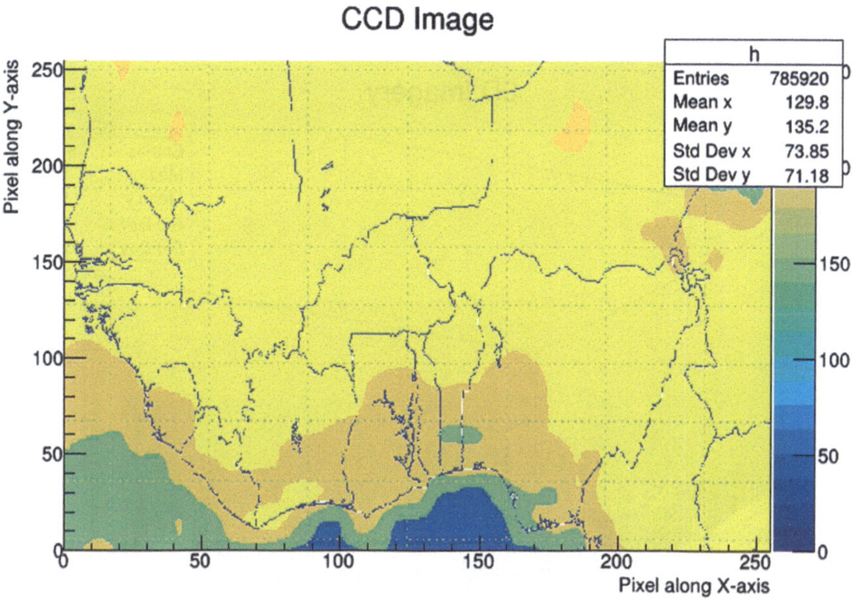

Script can be used to spot area of interest as shown below.

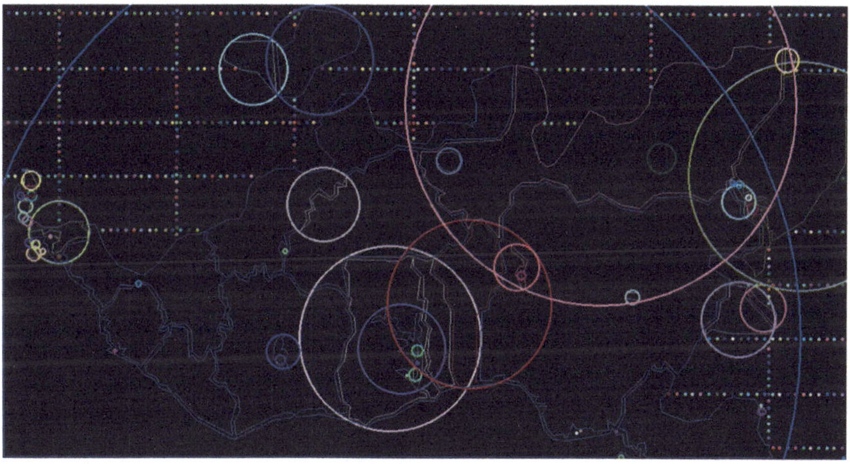

Script can be used to create a 3D image of the raw image as shown below.

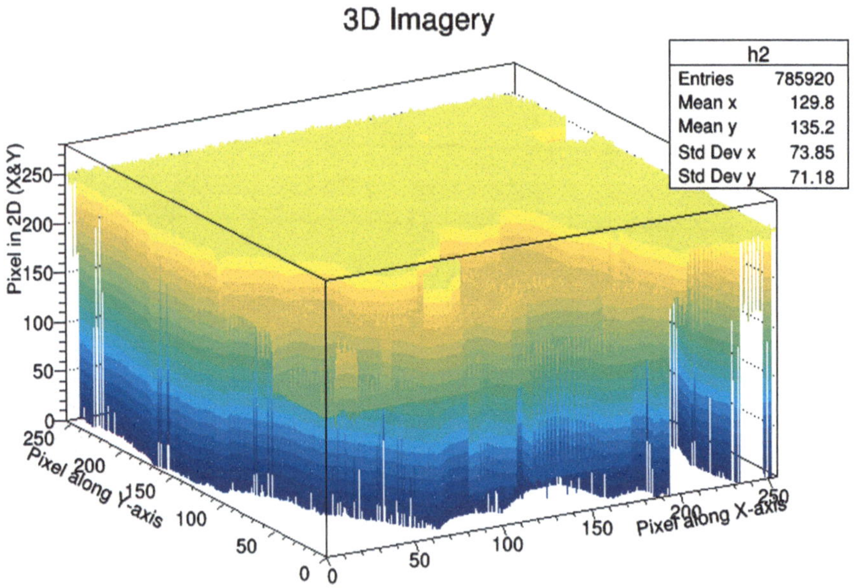

Script can be used to estimate the horizontal projections of the raw image as shown below.

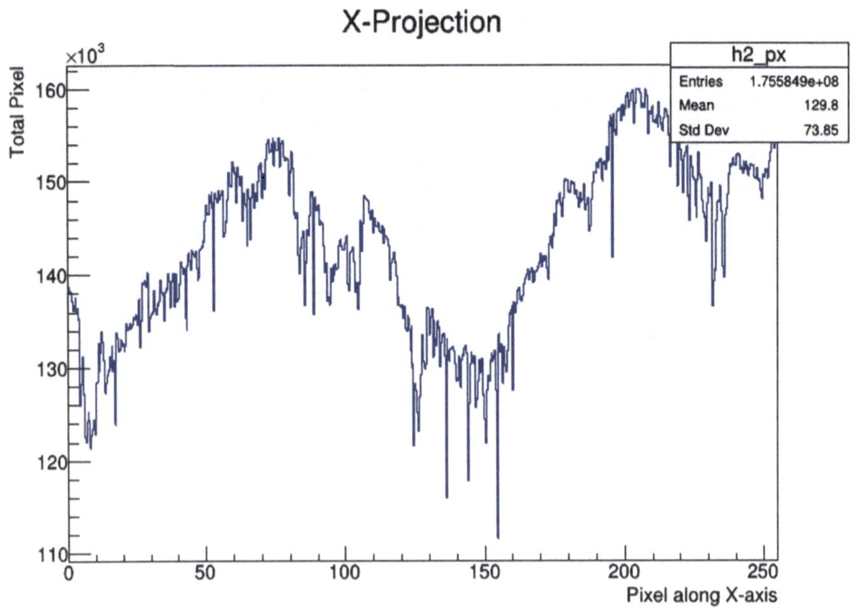

Script can be used to estimate the vertical projections of the raw image as shown below.

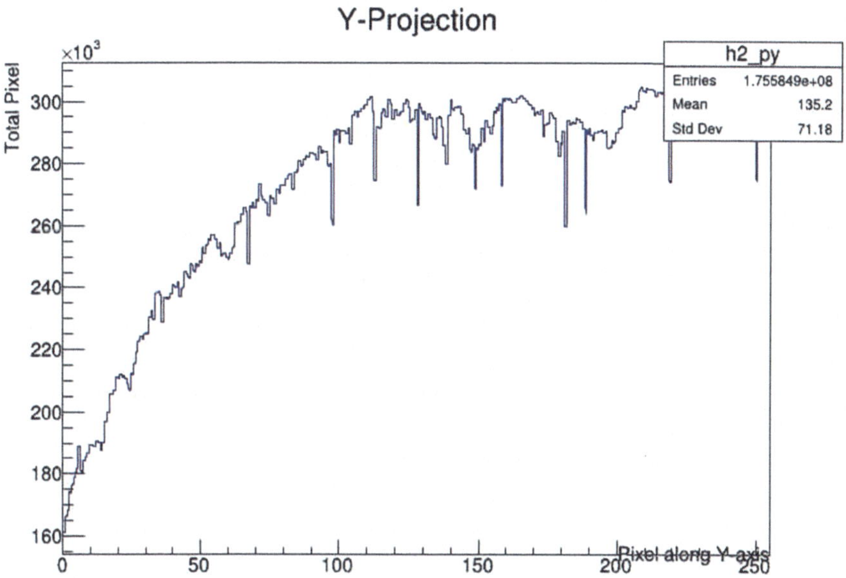

Script can be used to create a detailed 3D image of the raw image as shown below.

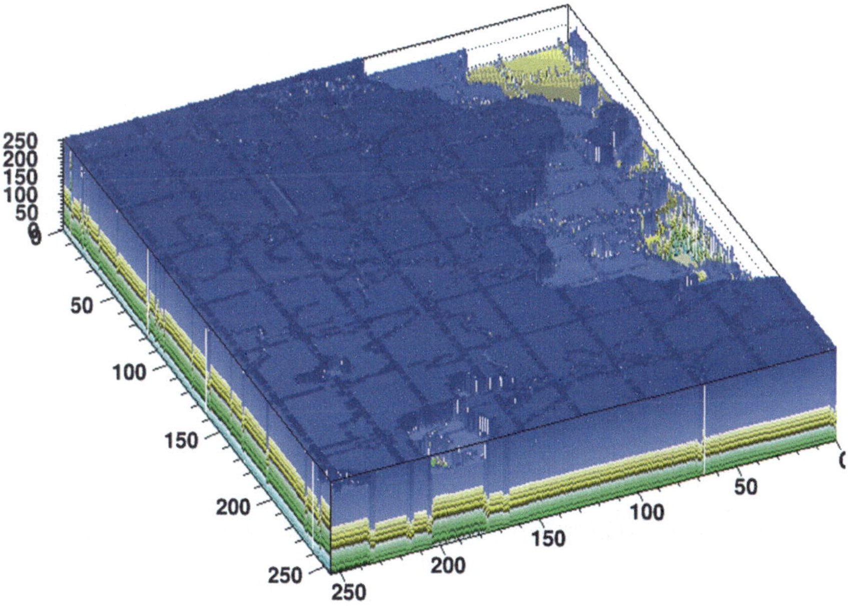

There is another side of image analysis that an environmental scientist can adopt. It is easier and friendly to use for field work. It is called the Google earth software. Environmentalist can view the field of research as shown below. It is a risk to go for field without examining the field using Google earth or other GIS software. The image obtained can be further analyzed using various software.

In the case of this work, Envi software was used to analyze the Google earth images (see screen shot below).

The 3D of the image can be obtained as presented in the diagram below.

As explained above, the Envi Software can be used to understand the Horizontal profile of the loaded image.

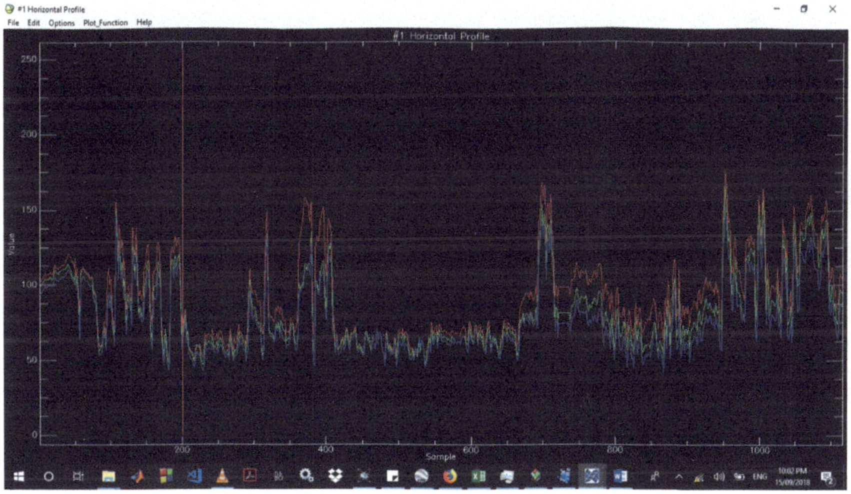

Likewise, the Envi Software can be used to understand the vertical profile of the loaded image. Typically, the horizontal and vertical profile are used to understand areas of special interest. For example, in the context of the Google earth image above, the regions of high peaks may refer to regions of geophysical interest.

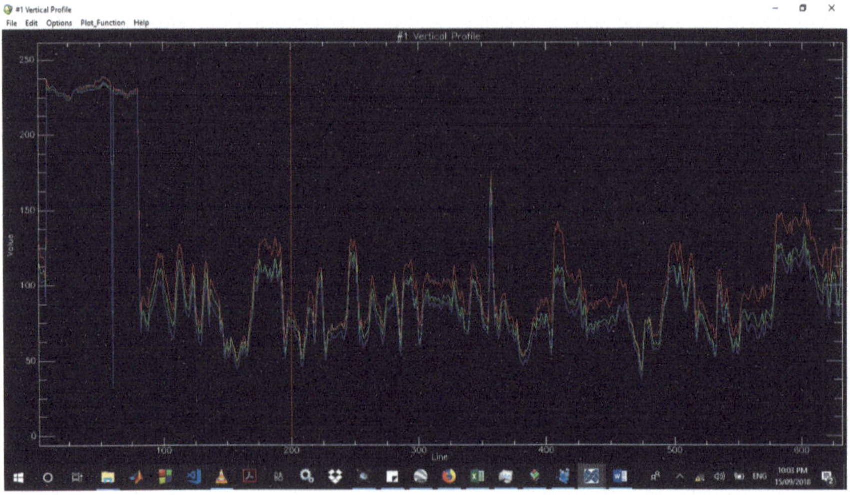

In same vein, the spectral profile of the Google earth image can be obtained as shown below.

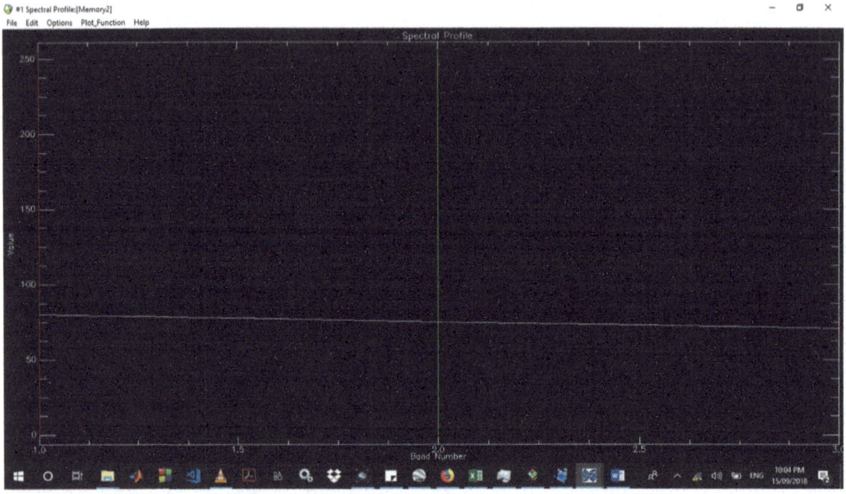

Also, the scattered plot of the Google earth image can be obtained as shown below.

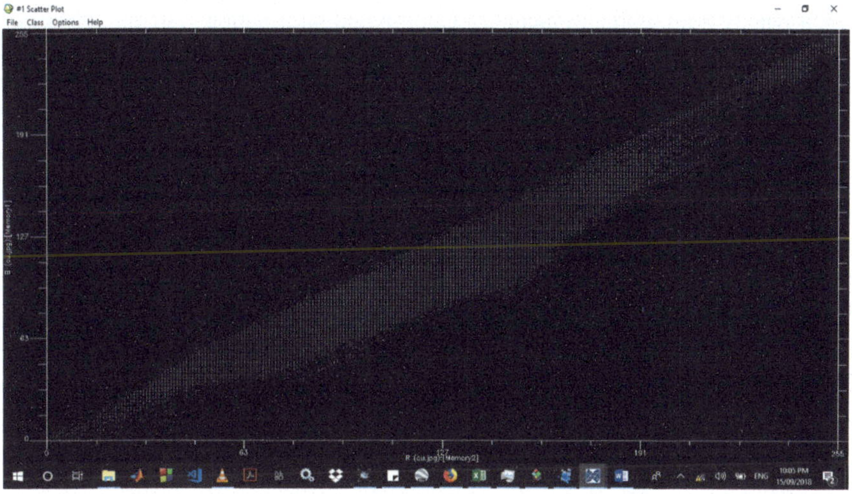

In other words, it has been clearly shown that analysis of image dataset is very essential in comprehending the interpretation of the features. Hence, the interpretation of image facially without using the software is not scientifically reasonable.

5.4 Special Cases and Comments

Geographic information system (GIS) is referred to a system that is programmed or created to import, interpret, analyze, store geographical data. Geographic Information Systems is a computer-based tool that analyzes, stores, manipulates and visualizes geographic information, usually in a map. In data creation, the common practice is digitization of information that are found on hard copy. For example, maps or survey plans are transferred into digital medium through the use of a CAD program (among other known techniques). Also digitization occurs in remote sensing to extract geographic data using ortho-rectified imagery. The digitized data are re-interpreted with respect to space-time (otherwise known as spatio-temporal allocation). Space-time concept can be with respect to dates/times of occurrence, coordinates (x, y, and z axis) longitude, latitude, or elevation. A digitizer produces vector data as an operator traces points, lines, and polygon boundaries from a map. The scanning of a map results in raster data that could be further processed to produce vector data. Hence, the process of inserting information into GIS is called data capture. However, it is mandatory that the data must be in digital form.

GIS is sometimes referred to as geographic information science (GIScience) i.e. the science underlying geographic concepts, applications, and systems. Other names synonymous with GIS include: "Geog-based information systems", "Natural resource information systems", "Geo-data systems", "Spatial information systems", "Geographic data systems", "Digital mapping systems", and "Land information systems" (LIS). The component of GIS are hardware, software, data and people. GIS can be applied to engineering, planning, management, transport/logistics, insurance, telecommunications, and business. The geographical data includes digital maps (like geographical, vegetative, aerial, satellite etc.), ubiquitous data (like data on earth quakes, hurricane, flood, air or water pollution etc.), Geospatial data, cartographic data, photographic data, digital data etc. Geospatial data is created, shared, and stored in two primary data types are raster and vector. Vector data is represented as either points, lines, or polygons. Vector data are data that have exact location, or hard boundaries. Raster data is referred to as continuous data. It has no exact location or hard boundaries. Rasters data are viewed as a series of grid cells where each cell has a value representing the feature being observed. For example, in a map grid, there are specific features that can be extracted or observed i.e. elevation, temperature, precipitation, or soil Ph. Hence, typical example of a raster data is remote sensing imagery, like aerial photography and satellite imagery.

The vector data has the following format: Shapefile (SHP) is an emerging format used in industries as standard geospatial data format that may occur with extensions shp = shapefile, shx = header and .dbf = associated database file; Smart Data Compression (SDC) is a compressed data format; Geodatabase (GDB) is a collection of geographic datasets of various types i.e. file, personal, and ArcSDE; ArcInfo Coverage has been phased out and is rarely seen/used today; Arc Export or Interchange Format (has extension '.e00') that are also rarely used today. Raster data has the following format: ArcInfo Grid which do not have an individual file extension but has folder such as "info" folder that typically contains .dat and .nit files; Band Interleaved by Line (BIL), Band Interleaved by Pixel (BIP), and Band Sequential (BSQ) that produce by remote sensing system as BIL, BIP, and BSQ respectively; Digital Elevation Model (DEM) is used by the USGS to record elevation information; GeoTIFF provides the Lat/Long boundary extent of the data; Light Detection and Ranging (LiDAR) is a remote sensing method that uses light in the form of a pulsed laser to measure ranges (variable distances) to the Earth with .LAS file extension. Aside the raster images and vector, there is a new hybrid method of storing data. The process entails the representation of real object as point clouds, which combine three-dimensional points with RGB information at each point to create a "3D colour image". Modern remote sensing techniques adopt this method. Also, some GIS applies this technique in a typical GIS thematic maps. This technique is now integrated in field work using hardware system such as GPS, laser rangefinder and computer to enhance accurate mapping and data analysis while on the field.

In GIS, topographical maps are the main source, and aerial photography and satellite imagery are extra sources for collecting data and identifying attributes which can be mapped in layers over a location facsimile of scale. The process of digitizing a map may be tedious because the map has to be checked within theoretical dimensions,

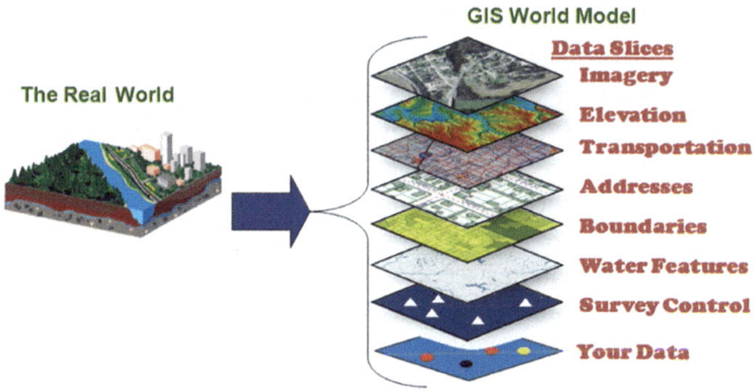

Fig. 5.5 GIS stratification of the real world [4]

and then scanned into a raster format, and resulting raster data has to be given a theoretical dimension by a rubber sheeting or warping technology process. This process is prone to mistakes because all geographical data are inherently inaccurate, and these inaccuracies will propagate through GIS operations in ways that are difficult to predict.

One of the uniqueness of the GIS is the analysis of locations with respect to coordinates. This information allows scientist appreciate patterns, trends, strata, features of the earth surface (Fig. 5.5).

The information generated on the GIS is a result of active integration of software and hardware. Most environmental scientists are drawn to the adoption of the GIS for in-depth interpretation of environmental trends, features and patterns. Sophisticated GIS software has the ability of splitting a geographical map into several incisive layers that have sensitive information. One of the advantages of GIS is its ability to visualize the problem and solution when it is eventually solved. In recent times, the GIS has grown tremendously as its application cuts-across multiple disciplines. The spatial emphasis of the GIS had made map interpretation and representation easier.

Most GIS software works on a supportive data aside the usual ubiquitous data. This supportive data is called the attribute data. Attribute data can be generally defined as additional information about each of the spatial features. This feature makes the GIS capable to do spatial analysis, manage large datasets, and display information in a map or in graphics. For example, the GIS software can used to find:

i. the location of features and relationships to other features;
ii. the spatial map of an object or location within the real-world;
iii. the visualization solutions to live-problems;
iv. the map of quantities, such as buildings etc.;
v. the density of features in a given space;
vi. what is happening nearby some feature or phenomenon;
vii. how a specific area has changed over time;
viii. extensive layer style control e.g. font/style/opacity settings for points/lines/areas/labels/legends/drawings;

ix. map that automatically lists displayed features and has a live scale bar;
x. the density or concentration of an event in a map;
xi. what is happening or what features are located inside a specific area/region;
xii. the features located inside a specific geographical enclave e.g. rocks, river, flood plain etc.;
xiii. thematic visualizations such as color, pattern/icon, dot-density, chart, scaled-symbol, and 3D prism themes;
xiv. what is happening within a set distance of a feature or event by mapping what is nearby using geoprocessing tools like BUFFER;
xv. stand-alone charting capabilities find pie, bar, line, area, scatter, and function charts;
xvi. anticipated future conditions via the study of changes over past years to decision or policy-making e.g. cases of incessant flooding in some part of the world;
xvii. the best geographic potential for a region to initiate renewable energy venture.

The characteristics of a good GIS database includes:

i. has maptitude GIS program that have a powerful proprietary relational database;
ii. has over 50 file types and more than 100 GIS and CAD formats;
iii. supports many formats such as dBase, CSV, ASCII, Lotus, Google KML, ArcGIS, MapInfo MIF, Oracle Spatial, SQL Server Spatial and AutoCAD DXF;
iv. easy data management;
v. allows model integration at any point;
vi. database can be modified as technology changes;
vii. database can be manipulated as one-to-one, one-to-many, or many-to-one joins
viii. has multiple filters per layer;
ix. allows the creation of topological/non-topological spatial databases;
x. can be integrated or linked to multiple files.

The operational mapping activities in GIS includes: Mineral Mapping, Agriculture Mapping, Pipeline Corridor Mapping, Defense Mapping, Airport Mapping, Land Cover Classification, Topographic Mapping, Urban Development, Pre and post 2D/3D seismic surveys, Environmental Impact Studies (EIS), Coastal erosion studies, Cadastre Mapping, Disaster Analysis etc.

In field work, digitization has been automated by the adoption of easy-to-use techniques such as: coordinate geometry (COGO) for inputting survey data; laser rangefinder to incorporate positions of dataset collected on the field; global navigation satellite system (GNSS) for importing maps into GIS; adoption of low-cost mapping-grade GPS units with decimeter accuracy in real time to avoid post process, import, and data update. Hence, aside that digitization aids the GIS processes, it is now a time saving tool in the process of creating and acquiring dataset. The process of acquiring data in GIS is very important. The remotely sensed data is becoming a major influence

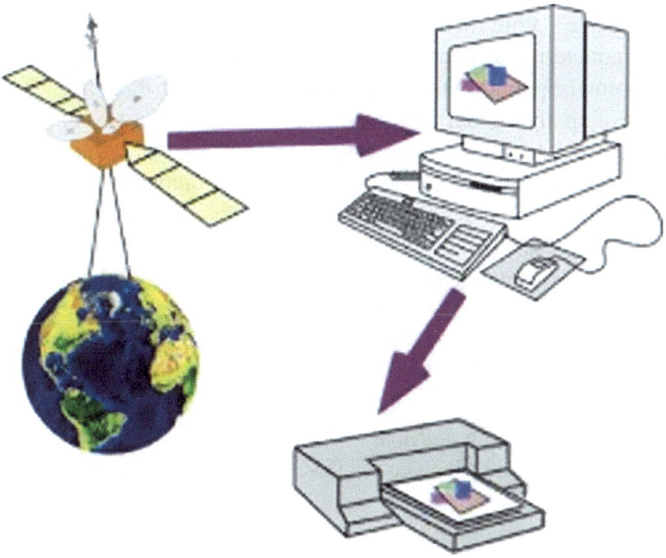

Fig. 5.6 The inclusion of remotely sensed data to GIS [5]

in GIS (Fig. 5.6) due to the sophisticated techniques such as adoption of cameras, digital scanners, helikites, miniature unmanned aerial vehicles (UAVs), radar and lidar. All this processes creates huge digitized data via photo interpretation of aerial photographs.

In other words, remote sensing techniques constitutes a major source for collecting raster data that can be further processed using different bands to identify objects and classes of interest, such as land cover, temperature, aerosol optical depth, sea surface temperature, snow cover, humidity, solar irradiance, altitudinal pressure, wind system, water bodies etc. After entering data into a GIS, the data usually requires editing, to remove errors, or further processing. The process to process data has been discussed in the previous topics under the title 'data treatment'. Before working on vector data, it must be made "topologically correct" before it can be used for some advanced analysis. Data restructuring is an important concept that environmental scientist must take note in recent times. This process allows the dataset to be readily useable for other processes such as the conversion of data into different formats in GIS operation. For example, a GIS may be used to convert a satellite image map to a vector structure by generating lines around all cells with the same classification, while determining the cell spatial relationships, such as adjacency or inclusion.

Aside, data treatment and restructuring, one important of GIS operation is GIS spatial analysis. This process allows for the use of analytical tools such as standard built-in facilities to automate information within the space-time concept. Spatial analysis has evolved into a new dimension in business intelligence known as "spatial intelligence". With this concept, GIS can be used to depict two- and three-dimensional characteristics of the Earth's surface, subsurface, and atmosphere from information points. Spatial analysis enhances topological modeling via the clear understanding of spatial model and geometric entities of the geographical area or region. Another aspect of spatial analysis is geometric networking which involves the interconnecting features. Geometric networking is composed of edges, which are connected at junction points to comprehend the various spatial locations. Geometric networks, are often represented by roads, rivers, and public utility grids in a GIS. Spatial analysis is also key to hydrological modeling because it helps to estimation processes to obtain variables such as slope, aspect and watershed or catchment area. In cartographic modeling, spatial analysis is used to obtain topographic layers such as forest-cover layer, geological layer etc. A new spatial analytic tool is the spatial ETL. It provide the data processing functionality of traditional extract, transform, load (ETL) software, but with a primary focus on the ability to manage spatial data. GIS users can translate data between different standards and proprietary formats using the spatial ETL.

Another special process in GIS operation is 'data extraction'. It is a GIS process similar to vector overlay, though it can be used in either vector or raster data analysis. It is used to extract the features of one data set that fall within the spatial extent of another dataset using "clip" or "mask". Data extraction has wide application in environmental science especially in fieldwork where huge data are obtained. This process is also common in remotely sensed dataset treatment. After doing data restructuring, data extraction is carried out to concentrate on a particular feature or area. Data extraction is used when GIS processes are used to register and combine the two images to produce three-dimensional perspective view.

The condition of the Earth's surface, atmosphere, and subsurface can be examined by feeding satellite data into a GIS (Fig. 5.7). GIS technology gives researchers the ability to examine the variations in Earth processes over days, months, and years. Most environmental researcher adopt this technique in their work. Hence, the total understanding of a given environmental data depends on the skills of the researcher to use GIS tools. In addition to the integration of time in environmental studies, GIS is also being explored for its ability to track and model the progress of human activities throughout their daily routines. Hence, GIS is found to be applicable not only in environmental field but in the fields of science, government, business, and industry, with applications including real estate, public health, mobile transport services e.g. Uber, Taxifier etc., crime mapping, business, social media, national defense, telecommunication, sustainable development, natural resources, banking, education, climatology, renewable energy planning, location-based services, landscape architecture, archaeology, regional and community planning, car tracking, airplane tracking, and logistics.

A Geographic Information System (GIS Software) is designed to store, retrieve, manage, display, and analyze all types of geographic and spatial data. GIS software lets you produce maps and other graphic displays of geographic information for analysis and presentation. There three main platforms of GIS softwares

Fig. 5.7 Sub-surface analysis using GIS (Intermountain [6])

i.e. SaaS—Software as a Service (ArcGIS, CartoDB, Mapbox), PaaS—Platform as a Service (Google Maps Javascript, Here Maps Javascript, Microsoft Bing, US Census Geocoder) and DaaS—Data as a Service (ArcGIS, Apple Maps, Here Maps, OpenStreetMap, Microsoft Bing Maps). The different GIS software includes: GRASS GIS, gvSIG, ILWIS (Integrated Land and Water Information System), JUMP GIS/OpenJUMP, MapWindow GIS, QGIS, SAGA GIS (System for Automated Geoscientific Analysis), uDig, Capaware, FalconView, Kalypso, TerraView, TerraLib, Whitebox GAT, GeoServer, MapGuide Open Source, Mapnik, MapServer, PostGIS, ArangoDB, TerraLib, GeoBase (Telogis GIS software), OpenLayers, Leafletjs, GeoTools, GDAL/OGR, Orfeo toolbox, GeoNetwork opensource, pycsw, Chameleon, MapPoint etc.

A simple illustration on how GIS works on remotely sensed data is presented in the screenshot shown in Fig. 5.8.

The technique of integrating models into GIS for the analysis of remotely sensed dataset are presented by Janssen et al. [7] as shown in Fig. 5.9. The processing unit requires the research skill to either program or use different GIS tools.

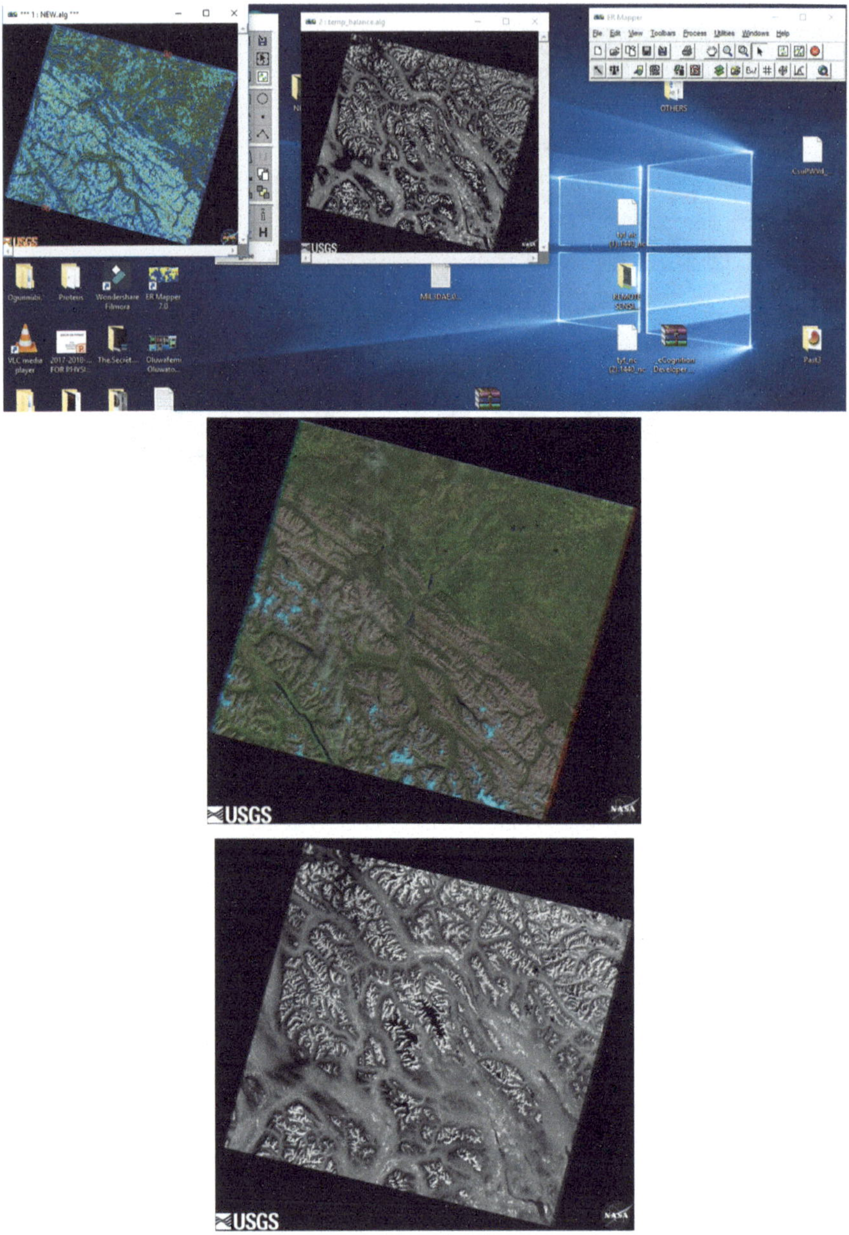

Fig. 5.8 Pictorial analysis of GIS images

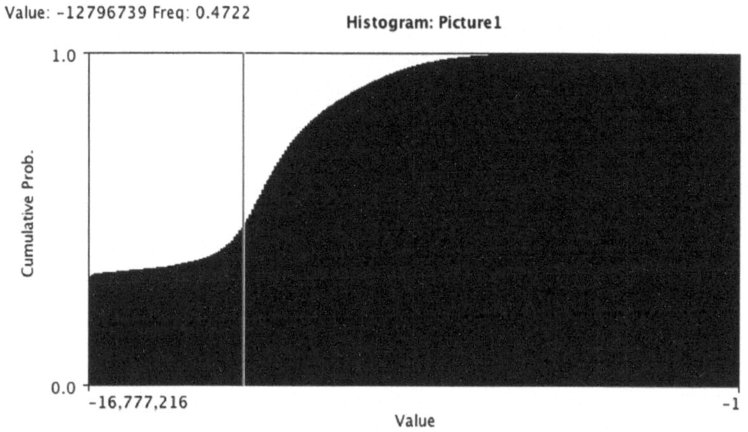

Fig. 5.8 (continued)

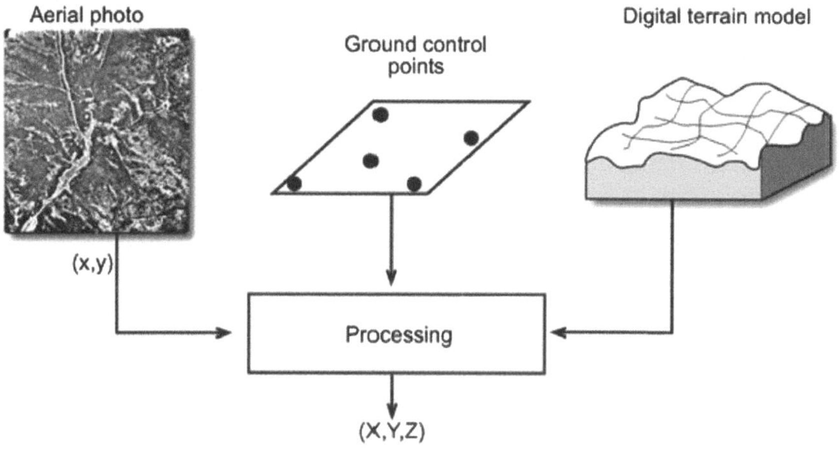

Fig. 5.9 Monoploting techniques [7]

It is advisable to automate the processes in Fig. 5.9 to reduce errors due to fatigue.

References

1. Kelvin S (2017) Public sector records management: A practical guide. https://books.google.com.ng/books?id=57UFDAAAQBAJ&printsec=frontcover&dq=inauthor:%22Kelvin+Smith%22&hl=en&sa=X&ved=0ahUKEwiq0e7nu8_mAhVHQxUIHdXVBO0QuwUIKzAA#v=onepage&q&f=false. Accessed 23 Apr 2019
2. Cheplaskei (2018) https://peda.net/kenya/css/subjects/computer-studies/form-three/driac2/data-processing/doeidp. Accessed 23 Apr 2019

3. Wes (2016) https://www.wes.org/privacy-policy/. Accessed 2 Feb 2019
4. Balch Springs Library-Learning Center (2018) https://www.balchspringslibrary.org/news-events/lib-cal/geographic-information-system-gis-mapping. Accessed 29 Aug 2018
5. Pidwirny M (2006) Introduction to geographic information systems. Fundamentals of physical geography, 2nd edn. http://www.physicalgeography.net/fundamentals/2f.html. Accessed 9 Sept 2018
6. Intermountain (2018) http://www.oilandgasbmps.org/resources/gis.php. Accessed 29 Aug 2018
7. Janssen LLF, Weir MJC, Grabmaier KA, Kerle N (2004) Geometric aspects. Kerle N, Janssen LLF, Huurneman GC (eds) Principles of remote sensing: an introductory textbook. ITC, The Netherlands, pp 143–158

Chapter 6
Conclusions

From the definition of environmental modeling, there is a lot to be known about the environment, its current challenges, sustainability and management. If the predictions of World Health Organization (WHO) on the number of deaths due to environmental disaster are anything to go by, then it behoves on environmental researcher to be very detailed in creating, acquiring, manipulating and analysing data. The reportage of environmental hazards and disaster must be properly synergized and situated for decision or policy maker to have a better understanding of the current challenges. The expansion of environmental discipline and research lines is a welcome-development; however, equipping research i.e. using various smart technologies is significantly important as creating fields of study. The principles of modeling in environmental study cut across intra-disciplinary and inter-disciplinary research lines. Hence, the first duty of beginner in the field is to identify what is the research trends in environmental studies? Where to get help? Which model are accurate in the field of interest? How to create or obtain necessary dataset?

For better illustration of what an environmental model looks like, a wide spectrum of old and emerging models was discussed. The model that was considered were air pollution models, water pollution models, hydrological environmental models and land pollution. It was clearly shown how environmental modeling has evolved from theoretical or academic exercise into easy-to-use computer model or software. The process for developing a good environmental model was highlighted in an orderly manner. The most vital of the developmental processes is the concept-development. Beginner in the field of environment most time have the challenge of identifying researchable topics that may likely to evolve into easy-to-use computer software. This is because there is stereotype research format in some environmental field that must be holistically reviewed. Hence, from the broad definition of individual environmental fields (given in the first chapter), a simpler way to identify research problem was illustrated by considering aerosol models and how they are formulated from concept to implementation stages.

M. E. Emetere and E. T. Akinlabi, *Introduction to Environmental Data Analysis and Modeling*, Lecture Notes in Networks and Systems 58,
https://doi.org/10.1007/978-3-030-36207-2_6

Environmental modeling is a large concept. Hence, the mode of interpretation for each field differ with respect to established experimental research or theoretical background. However, it was reported that some of the existing model are gradually becoming obsolete because of climate change or new discoveries. Hence, emerging aerosol models were discussed for readers to understand why and how they have evolved from the shortcoming that was seen with existing model. This aspect is very important for readers to appreciate also the existing shortcoming with the emerging models. Hence, the popular saying that no model is 100% correct may be right in its actual sense. It was also reported that for a model to be assumed almost accurate, the researcher should have research base that can help test the model or seek collaboration with organizations that have the funds to actualize the fronting of the model-globally.

Due to many years of data collection and the synergy of data sharing amongst institutions and organizations, environmental modeling has evolved into a more rigorous process via the advent of the big data concept. The five terms that defines if an environmental dataset can be classified as big data were discussed. The terms include volume, velocity, variety, variability and complexity. The full explanation of the big data concept signifies that remotely sensed data currently contributes the largest to body of environmental database. New environmental model must have capacity to accommodate huge dataset and summarize the analysis in precise numeric values or ranges. Hence, one of the parameter that must follow 'developing research concept' should be examining of past and present dataset with high precision on the importance of the research concept.

Numerical methods in environmental science were viewed from five perspectives i.e. physical models, models of natural science, mathematical model, engineering model and environmental model. Here, the subdivision of numerical methods was considered to solve specific problems. The methods include bisection method, Newton's method, secant method, Gaussian elimination method, interpolation method, iterative method, power method, numerical differentiation, numerical quadrature and numerical integration. An illustrative section was done by solving life problem i.e. using the aforementioned numerical methods. The accuracy, cumbersome nature, technicalities and usability of each numerical method were discussed via the examples presented. Also, the use of programming language (C++) to ease the estimation of numerical result was introduced.

An overview of the application of numerical analysis in environment studies or models was discussed holistically. First, the successes of numerical model software in environmental studies was highlighted in some branches of environmental research. Second, the need for environmental researcher to be in control of their analysis was emphasized. This involve patronizing reliable or proven software outlet to avoid reflection of systemic errors in the whole data analysis. This leads to wasted time, resources and efforts. Third, the parameter for developing an indigenous numerical model for solving specific environmental problems were highlighted. The main parameter for developing or executing numerical models is database. Database is a collection of dataset in whatever format i.e. asci, Binary NetCDF, HDF5. The list of databases and its mode of operation were listed for reader to appreciate the resources freely available online. The process for creating a personal reanalysis database using numerical techniques or methods was highlighted which includes: understanding the database you are working with; defining theoretical limits of the proposed forecast model; characterizing different scenarios in the project design; and applying the forecast model on known databases.

The computational techniques in environmental studies were introduced by presenting the basics of computational operation. This includes: knowing the steps to software creation; combining software with requisite hardware to achieve specific results; understanding the task of writing a program or scripts. The types of computer languages were reviewed holistically. Also, the different computer libraries were discussed to guide reader on how to choose and use a computer library when writing a script or program. Different types of computer techniques were discussed and how they can be integrated to perform a general or specific task.

The insights for novice or beginners in environmental field to step-up to professionals was reiterated by systematically highlighting the processes of developing a cost-free or low-cost computational technique for solving problems. The open source scientific packages were specifically discussed to educate reader on how to: use open source packages; integrate several open source packages; avoid known issues when using open source packages. Also, the open source libraries were discussed. The different open source libraries were highlighted. The compatibility issues of using individual open source library or combining open source libraries were holistically discussed.

The steps toward successful designing models that can work with huge data was discuss with salient illustrations on the process of data analytics to data execution. The sources of big data and the scope of big data analytics was highlighted and explained. The usefulness of big data analytics to small or big organization was discussed. Data sources were highlighted and explained to guide beginner or professional on where to obtain data related to different environmental projects.

The highlight of the research was the application of numerical methods to data analysis. The common pitfalls to avoid were listed for beginner. The four concepts in the application of numerical methods in data analysis was discussed and corresponding examples were shown to guide reader on how to set-up their own project. The use of computational technique was integrated into the examples to help reader choose the smart way of accomplishing a given task. This includes how to design C++ codes to analyze big data. However, the reader may adopt whatsoever computer language but the process remains the same all the same.

Data treatment has been a salient challenge in environmental studies. The ways of embarking on such task were highlighted and explained to help reader see its importance. Different ways of data treatment was illustrated with examples. The analogue and digital data processing techniques were discussed. One of the difficult aspect in environmental studies is the analysis of large dataset in form of images. The processes for applying numerical techniques to a huge dataset were illustrated with examples. The interpretation of some of the analysis was presented to help reader understand the nitty-gritty of modeling any type of dataset.

Lastly, special comment was drawn to the use of geographic information system (GIS) to boost the analytic skills of reader. The summarized mode of operation of the GIS was given. The need to apply the technique in environmental study was highlighted with examples.

This book gives novice or beginner a working manual that would help them measure their growth in whatever field of environmental science and engineering. The implementation of each topic is meant to create a medium for improving the authors initial concept. The professional or researcher are equipped with a training outline to raising beginner. Also, it enhances the advancement of different ideas presented in this book to aid environmental research work.

Appendix
Types of Dataset

Type of Data: CRUTEM3 Global mean temperature time series

Data format: '.dat'

Website: www.ipcc-data.org/documents/ar4_obs_gm/annual.land_ocean.90S.90N.df_1901-2000mean.dat

Date Accessed: 28th September, 2018

1880	−0.1591
1881	−0.0789
1882	−0.1313
1883	−0.1675
1884	−0.2485
1885	−0.2042
1886	−0.1695
1887	−0.2505
1888	−0.1605
1889	−0.1103
1890	−0.3014
1891	−0.2527
1892	−0.2926
1893	−0.3274
1894	−0.2880
1895	−0.2146
1896	−0.1009
1897	−0.1294
1898	−0.2517

(continued)

M. E. Emetere and E. T. Akinlabi, *Introduction to Environmental Data Analysis and Modeling*, Lecture Notes in Networks and Systems 58,
https://doi.org/10.1007/978-3-030-36207-2

(continued)

1899	−0.1356
1900	−0.1125
1901	−0.1693
1902	−0.2255
1903	−0.3442
1904	−0.3799
1905	−0.2583
1906	−0.2060
1907	−0.3726
1908	−0.3922
1909	−0.4041
1910	−0.3860
1911	−0.4144
1912	−0.3422
1913	−0.3196
1914	−0.1599
1915	−0.0810
1916	−0.2803
1917	−0.3227
1918	−0.2106
1919	−0.2186
1920	−0.2025
1921	−0.1390
1922	−0.2222
1923	−0.2063
1924	−0.1843
1925	−0.1260
1926	−0.0192
1927	−0.1007
1928	−0.0935
1929	−0.2201
1930	−0.0282
1931	0.0041
1932	−0.0285
1933	−0.1616
1934	−0.0285
1935	−0.0565
1936	−0.0318

(continued)

(continued)

1937	0.0727
1938	0.0955
1939	0.0898
1940	0.1234
1941	0.1837
1942	0.1377
1943	0.1299
1944	0.2362
1945	0.1162
1946	−0.0202
1947	−0.0406
1948	−0.0494
1949	−0.0752
1950	−0.1635
1951	−0.0029
1952	0.0353
1953	0.1190
1954	−0.1028
1955	−0.1220
1956	−0.1836
1957	0.0570
1958	0.1054
1959	0.0634
1960	0.0114
1961	0.0894
1962	0.1136
1963	0.1333
1964	−0.1268
1965	−0.0695
1966	−0.0117
1967	0.0022
1968	−0.0121
1969	0.0911
1970	0.0471
1971	−0.0550
1972	0.0348
1973	0.1617
1974	−0.0862

(continued)

(continued)

1975	−0.0177
1976	−0.1062
1977	0.1425
1978	0.0615
1979	0.1533
1980	0.1930
1981	0.2284
1982	0.1171
1983	0.2817
1984	0.0875
1985	0.0681
1986	0.1605
1987	0.2968
1988	0.3006
1989	0.2223
1990	0.3861
1991	0.3360
1992	0.2023
1993	0.2307
1994	0.2934
1995	0.4073
1996	0.2753
1997	0.4782
1998	0.5971
1999	0.4199
2000	0.3886
2001	0.5173
2002	0.5745
2003	0.5818
2004	0.5416
2005	0.6154
2006	0.5601
2007	0.5472
2008	0.4804
2009	0.5551
2010	−999.0000

Type of Data: NCDC Global mean temperature time series
Data format: ??
Website: www.ipcc-data.org/documents/ar4_obs_gm/annual.land_ocean.90S.90N.df_1901-2000mean.dat
Date Accessed: 28th September, 2018

1850	−1.757	−0.267	−0.409	−0.779	−0.552	0.130	−0.652
1850	3	3	3	3	2	2	3
1851	0.570	0.269	−0.711	0.402	−0.357	−0.148	−0.009
1851	3	3	3	3	3	3	3
1852	−0.226	0.166	−0.015	−0.979	−0.098	−0.140	0.029
1852	3	3	3	3	3	3	3
1853	0.055	−0.922	−1.433	−0.650	−0.325	−0.363	0.320
1853	3	3	3	3	3	3	3
1854	−0.254	0.340	−0.021	−0.044	0.489	0.034	0.577
1854	3	3	3	3	3	3	3
1855	0.328	−1.180	−0.476	0.530	0.337	0.219	0.309
1855	3	3	3	3	3	3	3
1856	−0.415	−0.580	−0.721	0.325	−0.424	0.276	−0.356
1856	4	4	4	4	4	4	4
1857	−1.054	0.450	−0.525	−0.961	−0.789	−0.051	0.129
1857	4	4	4	4	4	4	4
1858	0.320	−0.999	−0.450	−0.261	−0.306	0.141	−0.464
1858	4	4	4	4	4	4	4
1859	−0.267	−0.170	−0.212	−0.112	−0.034	−0.130	−0.121
1859	4	4	4	4	4	4	4
1860	0.340	−0.737	−0.956	−0.355	−0.291	0.106	−0.124
1860	4	4	4	4	4	4	4
1861	−1.247	0.203	0.180	−0.364	−0.820	0.245	−0.305
1861	4	4	4	4	4	4	4
1862	−0.894	−1.166	−0.230	−0.299	−0.208	−0.286	−0.533
1862	4	4	4	4	4	4	4
1863	0.827	0.195	−0.481	−0.360	−0.346	−0.427	−0.459
1863	3	4	4	4	4	3	3
1864	−1.278	−0.728	−0.539	−0.610	−0.466	−0.031	−0.117
1864	4	4	4	4	4	4	4
1865	−0.335	−1.076	−1.002	0.056	−0.212	0.019	0.043
1865	4	4	4	4	4	4	4
1866	0.279	−0.385	−0.801	−0.144	−0.471	−0.135	−0.233
1866	4	4	4	4	4	4	4
1867	−0.542	0.440	−1.428	−0.354	−0.755	−0.159	−0.369
1867	4	4	4	4	4	4	4

(continued)

(continued)

1868	−1.251	−0.852	−0.131	−0.589	0.211	0.066	0.238
1868	4	4	4	4	4	4	4
1869	0.062	0.464	−0.909	−0.298	−0.387	−0.252	−0.335
1869	5	5	5	5	5	5	5
1870	0.295	−0.432	−1.004	−0.313	0.011	0.394	0.093
1870	5	5	5	5	5	5	5
1871	−0.642	−0.841	−0.119	−0.236	−0.304	−0.180	0.102
1871	5	5	5	5	5	5	5
1872	0.134	−0.248	−0.363	−0.112	−0.058	0.103	−0.027
1872	5	5	5	5	5	5	5
1873	−0.088	−0.635	−0.466	−0.836	−0.265	0.514	−0.220
1873	6	6	6	5	6	6	6
1874	0.188	−0.617	−0.943	−0.392	−0.820	−0.253	−0.207
1874	6	6	6	6	6	6	6
1875	−0.894	−1.166	−1.239	−0.636	−0.177	−0.179	−0.216
1875	6	6	6	6	6	6	6
1876	−0.069	0.004	−0.222	−0.160	−0.346	−0.090	0.241
1876	7	7	7	7	7	7	7
1877	−0.003	0.198	−0.370	−0.461	−0.795	0.018	0.409
1877	6	7	7	7	7	7	7
1878	−0.091	0.293	0.669	0.510	−0.407	−0.130	0.112
1878	7	7	7	7	7	7	7
1879	−0.339	−0.381	−0.343	−0.438	−0.493	−0.507	−0.212
1879	7	7	7	7	7	7	7
1880	0.258	−0.096	−0.277	−0.162	−0.055	0.061	−0.089
1880	8	8	7	8	8	8	8
1881	−1.075	−0.447	−0.565	−0.338	0.074	−0.537	−0.300
1881	8	8	8	8	8	8	8
1882	0.722	0.182	0.314	−0.467	−0.571	−0.530	−0.395
1882	8	8	8	9	9	8	9
1883	−0.714	−0.688	−0.747	−0.517	−0.447	0.287	−0.164
1883	8	8	8	9	9	8	9
1884	−0.539	−0.341	−0.593	−0.876	−0.942	−0.629	−0.600
1884	9	9	9	9	9	9	9
1885	−1.039	−0.676	−0.883	−0.814	−0.718	−0.609	−0.350
1885	9	9	9	9	9	9	9
1886	−0.541	−0.794	−0.640	−0.137	−0.356	−0.552	−0.279
1886	9	9	9	9	9	9	9
1887	−0.428	−0.744	−0.436	−0.547	−0.593	−0.394	−0.306
1887	9	10	9	9	9	9	9
1888	−0.938	−0.889	−1.328	−0.231	−0.733	−0.291	−0.337
1888	10	10	10	10	10	10	10

(continued)

(continued)

1889	−0.318	−0.275	−0.123	0.023	0.123	−0.192	−0.330
1889	10	10	10	10	10	10	10
1890	−0.098	−0.467	−0.490	−0.238	−0.649	−0.231	−0.604
1890	11	11	11	11	11	11	11
1891	−0.990	−0.969	−0.725	−0.720	−0.404	−0.435	−0.590
1891	11	11	11	11	11	11	11
1892	−0.658	−0.026	−0.590	−0.831	−0.838	−0.483	−0.506
1892	11	11	11	11	11	11	11
1893	−1.791	−1.362	−0.627	−0.706	−0.747	−0.646	−0.206
1893	11	11	11	11	12	11	11
1894	−0.506	−0.566	−0.333	−0.532	−0.740	−0.695	−0.580
1894	12	12	12	12	12	12	12
1895	−1.036	−1.381	−0.745	−0.499	−0.530	−0.331	−0.786
1895	12	12	12	12	12	12	12
1896	−0.249	−0.342	−0.711	−0.722	−0.461	−0.389	−0.311
1896	12	12	12	12	12	12	12
1897	−0.586	−0.352	−0.692	0.035	−0.175	−0.070	−0.162
1897	13	12	13	13	13	13	13
1898	0.215	−0.427	−1.086	−0.671	−0.776	−0.168	−0.372
1898	13	13	13	13	13	13	13
1899	−0.311	−0.607	−0.619	−0.192	−0.519	−0.580	−0.305
1899	13	13	13	13	13	13	13
1900	−0.357	−0.233	−0.421	−0.370	−0.268	0.077	−0.254
1900	13	13	13	13	13	13	13
1901	−0.246	−0.454	−0.266	−0.166	−0.138	0.066	−0.108
1901	14	14	14	14	14	14	14
1902	−0.068	−0.085	−0.253	−0.344	−0.299	−0.217	−0.380
1902	14	14	14	14	14	14	14
1903	−0.142	0.046	−0.120	−0.541	−0.550	−0.537	−0.476
1903	14	14	14	14	14	14	14
1904	−0.863	−0.800	−0.923	−0.439	−0.547	−0.460	−0.477
1904	14	14	14	14	14	14	14
1905	−0.558	−1.203	−0.504	−0.613	−0.408	−0.153	−0.396
1905	15	15	15	15	15	15	15
1906	0.108	−0.313	−0.501	0.135	0.002	−0.144	−0.185
1906	15	15	15	15	15	15	15
1907	−0.636	−0.699	−0.546	−0.869	−0.922	−0.666	−0.496
1907	16	16	16	16	16	16	16
1908	−0.205	−0.406	−0.746	−0.543	−0.445	−0.399	−0.396
1908	16	16	16	16	16	16	16
1909	−0.623	−0.703	−0.737	−0.770	−0.802	−0.322	−0.395
1909	16	16	16	16	16	16	16

(continued)

(continued)

1910	−0.168	−0.396	−0.395	−0.255	−0.304	−0.200	−0.202
1910	16	16	16	16	17	16	16
1911	−0.603	−0.898	−0.752	−0.723	−0.468	−0.555	−0.299
1911	17	17	17	17	17	17	17
1912	−0.312	0.015	−0.391	−0.357	−0.310	−0.081	−0.586
1912	17	17	17	17	17	17	17
1913	−0.258	−0.576	−0.642	−0.218	−0.842	−0.631	−0.221
1913	17	17	17	17	17	17	17
1914	0.436	0.140	−0.101	−0.222	−0.082	0.009	−0.136
1914	17	17	17	17	17	17	17
1915	−0.107	0.394	−0.477	0.106	−0.299	−0.235	0.105
1915	17	17	17	17	17	17	17
1916	0.028	−0.054	−0.562	−0.309	−0.310	−0.518	−0.296
1916	17	17	17	17	17	17	17
1917	−0.573	−1.247	−1.118	−0.773	−1.203	−0.420	−0.104
1917	17	17	17	17	17	17	17
1918	−0.886	−0.673	−0.617	−0.627	−0.781	−0.351	−0.521
1918	17	17	17	17	17	17	17
1919	−0.034	−0.202	−0.422	0.009	−0.291	−0.049	−0.212
1919	17	17	17	17	17	17	17
1920	−0.171	−0.446	−0.201	−0.350	−0.258	−0.263	−0.178
1920	16	17	16	16	17	17	17
1921	0.150	−0.195	−0.131	−0.134	0.090	0.058	−0.062
1921	17	17	17	17	18	17	18
1922	−0.553	−0.412	−0.329	−0.022	−0.219	−0.241	−0.267
1922	18	18	18	18	18	18	18
1923	0.012	−0.479	−0.277	−0.611	−0.375	−0.252	−0.572
1923	18	18	18	18	18	18	18
1924	−0.434	−0.506	−0.472	−0.681	−0.399	−0.174	−0.215
1924	19	19	19	19	19	18	19
1925	−0.521	−0.176	−0.295	−0.176	−0.444	−0.325	−0.472
1925	19	19	19	19	19	19	19
1926	0.328	0.552	0.102	−0.234	−0.561	−0.118	−0.211
1926	19	19	20	20	20	20	20
1927	−0.205	−0.171	−0.324	−0.407	−0.505	−0.246	−0.116
1927	20	20	20	20	20	20	20
1928	−0.041	−0.176	−0.399	−0.132	−0.386	−0.542	−0.112
1928	20	20	20	20	20	20	20
1929	−0.517	−1.046	−0.500	−0.618	−0.504	−0.393	−0.443
1929	20	20	20	20	20	20	20
1930	−0.392	−0.085	−0.083	−0.165	−0.287	0.040	−0.041
1930	20	20	20	20	20	20	20

(continued)

(continued)

1931	−0.001	−0.303	−0.228	−0.392	−0.355	−0.015	0.034
1931	21	21	21	21	21	21	21
1932	0.761	−0.347	−0.484	0.041	−0.231	−0.192	−0.021
1932	21	21	21	21	21	21	21
1933	−0.396	−0.478	−0.371	−0.252	−0.197	−0.186	−0.153
1933	21	21	22	22	22	22	22
1934	−0.105	−0.082	−0.339	−0.379	0.161	0.020	0.056
1934	22	22	22	22	22	22	22
1935	−0.378	0.520	−0.021	−0.564	−0.443	−0.182	−0.066
1935	22	22	22	22	22	22	22
1936	−0.213	−0.708	−0.303	−0.330	−0.168	−0.082	0.292
1936	22	23	23	23	23	23	23
1937	−0.296	0.044	−0.539	−0.359	−0.117	0.060	−0.069
1937	23	23	23	23	23	23	23
1938	0.016	0.084	0.305	0.241	0.113	−0.030	0.063
1938	23	23	23	23	23	23	23
1939	0.160	−0.019	−0.481	−0.236	0.103	0.149	−0.062
1939	23	24	23	24	24	23	24
1940	−0.624	−0.131	−0.080	−0.111	−0.198	0.188	0.158
1940	23	23	23	23	23	23	23
1941	−0.161	0.095	−0.241	0.046	−0.033	0.144	0.204
1941	25	25	25	25	25	25	25
1942	0.115	−0.291	−0.092	−0.044	−0.106	0.078	−0.136
1942	25	25	25	25	25	25	25
1943	−0.352	0.109	−0.280	−0.005	−0.075	−0.299	−0.005
1943	26	26	26	26	25	26	25
1944	0.505	0.147	−0.021	−0.170	−0.163	−0.006	0.120
1944	25	25	26	25	25	25	25
1945	−0.205	−0.379	−0.143	0.151	−0.362	−0.141	−0.207
1945	26	26	26	25	25	26	26
1946	0.029	0.143	−0.289	0.076	−0.055	−0.285	0.013
1946	26	26	26	26	26	26	26
1947	−0.155	−0.178	−0.006	0.047	−0.016	0.147	−0.054
1947	27	27	27	27	27	27	27
1948	0.256	−0.127	−0.469	−0.125	0.056	0.127	−0.108
1948	27	27	27	27	27	27	27
1949	0.212	−0.416	−0.383	−0.237	−0.174	−0.226	−0.112
1949	28	28	28	28	28	28	28
1950	−0.608	−0.376	−0.257	−0.297	−0.187	−0.098	−0.200
1950	29	29	29	29	29	29	29
1951	−0.537	−0.734	−0.328	−0.182	−0.068	−0.075	−0.060
1951	33	34	33	33	34	33	34

(continued)

(continued)

1952	0.299	0.083	−0.246	−0.063	−0.045	−0.082	0.078
1952	34	34	34	34	34	34	34
1953	0.077	0.109	0.183	0.123	−0.016	0.098	−0.150
1953	34	34	34	35	34	35	34
1954	−0.457	−0.202	−0.272	−0.283	−0.322	−0.191	−0.238
1954	35	35	35	35	35	35	35
1955	0.243	−0.088	−0.648	−0.378	−0.262	−0.141	−0.299
1955	35	35	35	35	35	35	35
1956	−0.432	−0.669	−0.541	−0.488	−0.483	−0.308	−0.259
1956	36	36	36	36	36	36	36
1957	−0.285	−0.234	−0.320	−0.130	−0.073	0.064	−0.239
1957	36	36	36	36	36	36	36
1958	0.424	0.256	0.080	0.126	0.085	−0.037	0.085
1958	36	36	36	36	36	36	36
1959	0.071	0.045	0.154	0.027	−0.186	0.083	0.062
1959	36	36	36	36	36	36	36
1960	0.057	0.317	−0.610	−0.324	−0.449	−0.026	−0.172
1960	36	36	36	36	36	36	36
1961	0.068	0.207	0.057	0.090	0.097	0.067	−0.039
1961	36	36	37	36	37	37	37
1962	0.077	0.163	−0.036	−0.072	−0.163	−0.068	−0.123
1962	37	37	37	36	37	36	37
1963	−0.152	0.314	−0.231	−0.214	−0.152	−0.103	0.090
1963	37	37	37	37	37	37	37
1964	0.001	−0.293	−0.382	−0.279	−0.210	−0.297	−0.218
1964	37	37	37	37	37	37	37
1965	−0.113	−0.311	−0.293	−0.463	−0.202	−0.033	−0.310
1965	37	37	37	37	37	37	37
1966	−0.148	−0.058	−0.031	−0.197	−0.189	0.098	−0.002
1966	37	37	37	37	37	37	37
1967	−0.194	−0.403	−0.220	−0.074	0.102	−0.278	−0.172
1967	37	37	38	37	37	37	37
1968	−0.331	−0.311	0.195	−0.283	−0.402	−0.290	−0.151
1968	37	37	37	37	37	37	37
1969	−0.454	−0.464	−0.142	0.068	0.035	−0.042	0.056
1969	38	37	37	37	38	37	37
1970	−0.019	0.336	−0.181	0.086	−0.082	0.144	0.001
1970	37	37	37	37	37	37	37
1971	−0.104	−0.404	−0.389	−0.353	−0.342	−0.345	−0.130
1971	36	36	36	36	36	36	36
1972	−0.671	−0.617	−0.312	−0.197	−0.133	0.045	−0.043
1972	36	36	36	36	36	36	35

(continued)

(continued)

1973	0.300	0.456	0.408	0.249	0.190	0.306	0.106
1973	35	35	35	36	36	35	35
1974	−0.475	−0.604	−0.257	−0.231	−0.252	−0.213	−0.069
1974	35	36	36	36	36	36	35
1975	−0.003	−0.098	−0.061	−0.047	0.011	−0.013	−0.034
1975	35	35	35	35	35	35	35
1976	−0.227	−0.394	−0.738	−0.299	−0.352	−0.307	−0.282
1976	35	35	35	35	35	35	35
1977	−0.373	0.161	0.282	0.202	0.037	0.173	0.067
1977	35	35	35	35	35	35	34
1978	0.025	0.068	0.139	0.009	−0.049	−0.177	−0.026
1978	35	35	35	35	35	35	35
1979	−0.057	−0.193	0.089	−0.170	−0.112	0.080	0.006
1979	34	34	35	34	34	34	34
1980	0.097	0.195	0.021	0.220	0.315	0.118	0.044
1980	34	34	34	34	34	34	34
1981	0.550	0.493	0.449	0.388	0.147	0.093	0.117
1981	35	36	36	35	35	35	35
1982	−0.051	0.050	−0.172	0.000	0.070	−0.115	−0.004
1982	34	34	34	34	34	35	35
1983	0.640	0.543	0.360	0.208	0.242	0.123	0.204
1983	34	35	35	35	35	35	35
1984	0.015	−0.075	−0.009	−0.093	0.190	−0.031	−0.051
1984	35	35	35	35	35	35	35
1985	0.070	−0.227	−0.015	0.020	0.073	−0.054	−0.085
1985	35	35	35	36	35	35	35
1986	0.389	0.183	0.212	0.296	0.210	0.162	−0.039
1986	35	35	35	35	35	34	35
1987	0.259	0.642	−0.045	0.301	0.146	0.262	0.387
1987	35	35	35	35	35	34	35
1988	0.595	0.254	0.438	0.340	0.317	0.390	0.291
1988	35	35	35	35	35	34	34
1989	0.142	0.322	0.256	0.210	0.117	0.119	0.158
1989	34	35	34	34	34	34	35
1990	0.429	0.414	0.910	0.452	0.339	0.347	0.270
1990	34	34	34	34	33	33	33
1991	0.371	0.523	0.330	0.508	0.316	0.554	0.394
1991	32	31	31	31	30	31	31
1992	0.559	0.416	0.447	0.150	0.167	0.165	−0.212
1992	31	30	30	30	31	31	30
1993	0.362	0.219	0.295	0.266	0.286	0.198	0.203
1993	31	31	30	31	30	30	31

(continued)

(continued)

1994	0.371	−0.166	0.259	0.386	0.436	0.419	0.267
1994	31	31	31	31	30	31	31
1995	0.682	0.810	0.360	0.344	0.293	0.475	0.435
1995	31	30	31	31	31	31	30
1996	0.145	0.319	0.090	−0.003	0.265	0.272	0.261
1996	31	30	31	30	30	30	30
1997	0.346	0.526	0.300	0.201	0.277	0.530	0.453
1997	30	30	30	30	30	30	30
1998	0.726	1.369	0.779	0.956	0.873	0.850	0.928
1998	30	31	30	30	30	31	31
1999	0.697	0.946	0.353	0.354	0.394	0.440	0.450
1999	30	31	30	30	30	31	31
2000	0.218	0.659	0.529	0.705	0.327	0.345	0.288
2000	30	30	30	30	30	30	30
2001	0.611	0.499	0.660	0.523	0.524	0.469	0.523
2001	27	28	28	28	27	27	27
2002	0.934	1.000	0.898	0.686	0.595	0.513	0.685
2002	27	27	28	28	28	28	28
2003	0.811	0.573	0.448	0.532	0.749	0.774	0.527
2003	28	28	27	28	28	28	28
2004	0.730	0.906	0.822	0.714	0.228	0.539	0.329
2004	29	28	28	28	27	27	28
2005	0.767	0.563	0.783	0.882	0.673	0.825	0.801
2005	28	27	27	27	28	27	27
2006	0.657	0.740	0.638	0.551	0.269	0.655	0.797
2006	26	27	27	27	27	27	27
2007	1.104	0.867	0.749	0.851	0.623	0.412	0.403
2007	27	27	26	26	26	28	27
2008	0.234	0.337	0.907	0.377	0.297	0.428	0.690
2008	27	26	26	28	27	27	26
2009	0.576	0.731	0.615	0.694	0.536	0.566	0.495
2009	28	29	28	27	28	28	28
2010	0.727	0.687	0.971	0.825	0.736	0.000	0.000
2010	27	26	27	26	26	0	0

Type of Data: Full Display of IL_SALINERIVER_2010_000002
Data format: ??
Website: https://earthexplorer.usgs.gov/metadata/10480/367697/
Date Accessed: 28th September, 2018

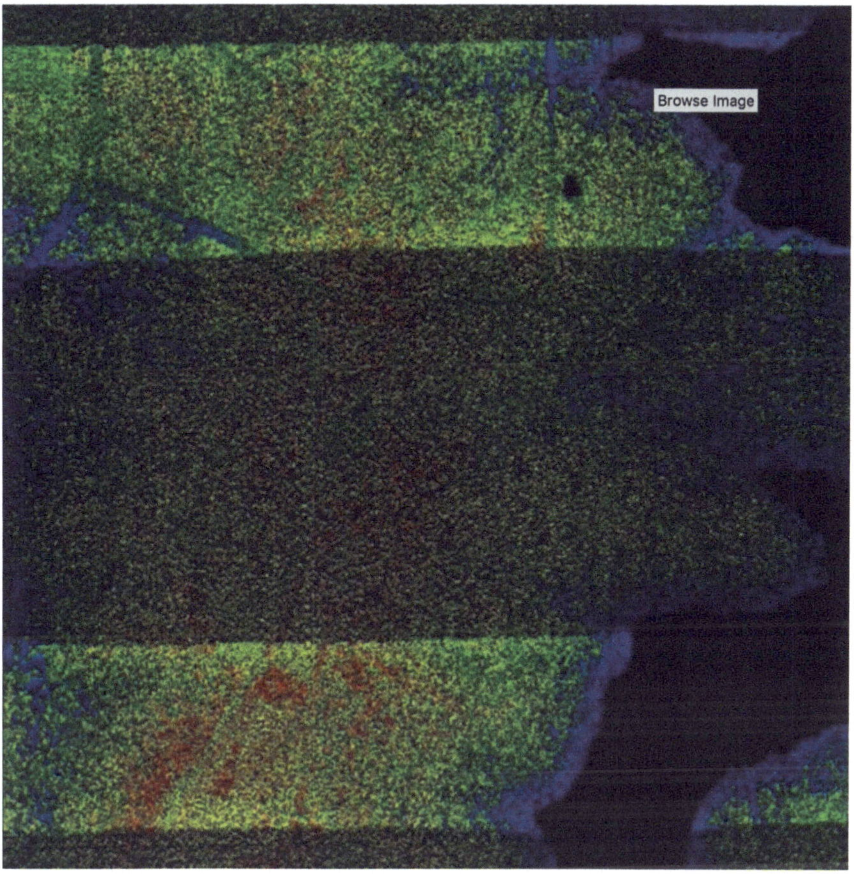

Data set attribute	Attribute value
LIDAR Entity ID	IL_SALINERIVER_2010_000002
Project	IL_SALINERIVER_2010
State/Province/Country	IL
Pulse and NPS Units	METER
Pulse Density	0.425266
Nominal Pulse Spacing	1.53345

(continued)

(continued)

Data set attribute	Attribute value
Beginning Date	01-SEP-10
Ending Date	01-DEC-10
Vendor	WOOLPERT_INC
Vendor Scene ID	16_309165
Map Projection	UTM
Projection Zone	16
Datum	NAD83
Vertical Datum	NAVD88
Center Latitude	37°37′25.30″N
Center Longitude	89°09′20.96″W
NW Corner Lat	37°37′49.06″N
NW Corner Long	89°09′52.24″W
NE Corner Lat	37°37′50.17″N
NE Corner Long	89°08′51,08″W
SE Corner Lat	37°37′01.54″N
SE Corner Long	89°08′49.68″W
SW Corner Lat	37°37′00.42″N
SW Corner Long	89°09′50.83″W
Center Latitude dec	37.6236943
Center Longitude dec	−89.1558222
NW Corner Lat dec	37.6302943
NW Corner Long dec	−89.1645111
NE Corner Lat dec	37.6306027
NE Corner Long dec	−89.1475221
SE Corner Lat dec	37.6170943

Bibliography

1. Austin Consultant (2018) https://www.austinconsultants.co.uk/solutions/big-data-analytics/. Accessed 29 Aug 2018
2. Badajena CK, Rout C (2012) Incorporating Hidden Markov model into anomaly detection technique for network intrusion detection. Int J Comput Appl 53(11):42–47
3. Guha-Sapir D, Hoyois P, Wallemacq P, Below R (2017) Annual disaster statistical review 2016: the numbers and trends. CRED, Brussels
4. Infotreesolutions (2018) http://www.infotreesolutions.com/services/big-data-services/. Accessed 29 Aug 2018
5. Korolkovas A (2016) Entangled polymer flows at interfaces. A thesis submitted to University of Upsala
6. Phillip (2016) https://medium.com/@philip4/big-data-analytics-trends-6a83e6bffb80. Accessed 29 Aug 2018
7. Sanni SE, Emetere M (2016) Mathematical modeling of insitu-bioremediation of crude oil polluted soil. Sci Eng Appl 1(4):27–32

M. E. Emetere and E. T. Akinlabi, *Introduction to Environmental Data Analysis and Modeling*, Lecture Notes in Networks and Systems 58,
https://doi.org/10.1007/978-3-030-36207-2

The manufacturer's authorised representative in the EU is Springer Nature Customer Service Centre GmbH, Europaplatz 3, 69115 Heidelberg, Germany. If you have any concerns regarding our products, please contact ProductSafety@springernature.com

Printed and bound by CPI Group (UK) Ltd, Croydon, CR0 4YY
30/09/2026
02225458-0007